sis
for
Profitable Business Applications

Systems Analysis for Profitable Business Applications

Ralph Cornes

PRENTICE HALL
New York London Toronto Sydney Tokyo

First published 1989 by
Prentice Hall International (UK) Ltd
66 Wood Lane End, Hemel Hempstead
Hertfordshire, HP2 4RG
A division of
Simon & Schuster International Group

Printed and bound in Great Britain by
A. Wheaton and Co. Ltd, Exeter

Library of Congress Cataloging-in-Publication Data

Cornes, Ralph.
Systems analysis for profitable business applications.
Bibliography: p.
Includes index.
1. Business–Data processing — Management.
2. Management information systems — Planning.
3. System analysis. I. Title
HF5548.2.C6727 1989 658'.05421 88–32491
ISBN 0–13–881319–1 (case)

British Library Cataloguing in Publication Data

Cornes, Ralph
Systems analysis for profitable business applications.
1. Data processing. Systems analysis
I. Title
004.2'1

ISBN 0–13–881319–1
ISBN 0–13–881327–2 pbk

1 2 3 4 5 93 92 91 90 89

ISBN 0-13-881319-1
ISBN 0-13-881327-2 PBK

Contents

Preface

The bad old days of the sixties are fortunately past, when the introduction of a major computer application usually led to weeks of crisis and perhaps to the discovery of major operational shortcomings. Computer people are much more professional today and they have learned from experience. Their problem is that as their professionalism and productivity have increased so have the expectations of management and the end-user, both of whom still tend to be wary of major new applications.

Managers no longer complain that computer applications do not work, but rather that they do not work well enough. They find that they want information which is in the application somewhere but that the application is not designed to make it easily available. Or they find that applications cannot be integrated to provide a higher order of benefits. And always during the development process, if they are involved as they should be, they find that they are asked to think in a way which is distinctly nonmanagerial.

End-users usually complain that the application has snags and undesirable side-effects which were not foreseen during development. Their reaction to a new computer application is enthusiastic when it removes a major bugbear, but otherwise ranges from grudging acceptance to passive resistance in the worst cases. The previous system has to be very poor before a new computer application is actively welcomed.

Given the power and reliability of modern hardware and software and the high level of competence and common sense of computer staff, this is a puzzling and an unsatisfactory state of affairs. The implicit theme of this book is that the major cause of computers failing to meet these enhanced expectations of managers and end-users lies in the way that major applications are developed.

Current development methods evolved to tackle the crises of the sixties and seventies; and they are still necessary because without them these crises would return. However, they should be enhanced to cope with today's problems, in many ways the problems of success. From the user's and manager's point of view, the wrong application is often constructed in the sense that they know it could do much more. The reason usually lies in the simple fact that managers, users and computer professionals find it difficult to get on the same wavelength during development.

This book addresses this as a key problem. It shows how the problem can be tackled and traces the implications of tackling it on the subsequent analysis and initial stages of the design process. It is meant to be recommended to line managers by DP professionals, and read by both.

It contains two new points for the DP professional. The first is that the application should be defined to a great level of detail before the computer content of the application is defined. The second is that the DP view of an entity (e.g. something in the real world about which information is gathered) should be refined to consider entities as either primary objects, or conceptual groups of objects, or activities which are the temporary association of two or three objects.

It contains some axioms about the nature of systems development which are always obvious and should be self-evident. Usually they have not been made in print in the past. If a statement seems self-evident, it is included to indicate that it is usually not treated in standard development methods. An example is the difference between DP and MIS applications, which every practical analyst knows about but, to my knowledge, is not mentioned in books on systems analysis. Another is the importance of using by-product input data.

As it happens, this approach simplifies the business of systems development. It is not opposed to the use of traditional structured documentation, which can continue in use where necessary. There is additional documentation suggested and in many cases this might replace some existing standards. But it is important to stress that this book is based on the same view of the development process as SSADM, Jackson, Yourdon *et al.* It is different to them only because it considers additional factors.

The individual techniques have been used by the author for years. They actually do work. It seemed time to bring them all together as this is one way to relate them to existing practice, and at the same time to indicate how applications could be developed which fit in better with the enhanced expectations of management and end-users.

Introduction

Whenever a new application is developed it involves a compromise between what is needed operationally and what is technically feasible, a match is made between computer resources and operational requirements. Unfortunately in developing many computer applications this match is botched. In many cases a rough match is made between operational requirements and computer capabilities, and from that point on the analyst designs outwards from the computer and allows computer requirements to dictate the shape of the final system.

Validation for instance is more likely to be fixed by the features available in the data dictionary or by the facilities provided by a file of check-tables rather than by the likely error rate of the terminal operator. Screens are likely to be designed as variations of existing screens rather than based on an analysis of the source documents which carry screen information. Program functions are likely to be finalized to simplify the computer programming rather than to provide the flexibility and functionality demanded by end-users.

There is nothing in existing development methods which states that this must happen, but equally there is nothing which says that this pitfall exists in the path of every business application. The theme of this book is that a change of emphasis from computer-oriented definitions to application-oriented definitions is necessary at every stage of the development process if the application is to meet user expectations.

One result of this approach might be to show that perhaps the provision of a computer service, or alternatively an end-user tool, might fit user needs better than the development of the traditional type of system. A spreadsheet with links to the accountancy books can often consolidate accounts more effectively and much more cheaply than purpose-written application programs.

A bonus to the developer is that the approach seems to simplify the development process. The massively complex documentation of most development methodologies is not rendered redundant, but the cases where it is needed in all its pomp are fewer.

The reason is that development documentation has many different purposes. It is meant to ensure consistency and completeness of design. It serves as a basis for communication between the team members. It is meant to provide an essential base for an evaluation and analysis of the application before installation, and to help with subsequent maintenance. And it is meant to aid communications with management and the end-user during development.

If at every stage reference is made back to the user-view of the application the need for documentation to communicate technical information to the user is avoided. Additionally, much of the work of ensuring design consistency and completeness is avoided if checks are made against the baseline established at the start of the project instead of against the documentation of the previous stage in the development process.

A consequence of this changed emphasis is that it becomes necessary to refine the idea of entity modelling (the way in which computer professionals define the world before they model it in an application). The change is necessary to cope with the way that managers and end-users think about their needs. But this too has a bonus, it simplifies the business of analysis and design. The refining of entity modelling can produce development procedures which are simpler and more effective versions of the procedures based on the traditional ideas of entities.

Entities are divided into primary objects (or things), group objects or concepts (customers in areas, or products in classes), and activities (temporary associations of objects as in a sale, or an order). The main reason is that managers and users think like this, but an important consequence is that the division allows a more systematic approach to data analysis.

In order to illustrate the procedures to be followed in developing business applications, the book has to provide examples of documentation. Documentation provides a framework of control for systems development activities. Its detailed format is unimportant as long as it provides this control. The documentation in this book should not be taken as holy writ, but as an example of the kind of documentation that should be produced to formalize the change of emphasis during development. It is oriented to providing management and user views of the application during development.

If SSADM, or LBMS, or Jackson, or Martin, or any other standards are in use and they serve to document the activities outlined in this book, there is every reason to continue to use them. They define and control the stages of systems development, and they can be used to control a user-oriented definition just as easily as they are used to control a computer-oriented definition. The additional documents that enforce the user-view of the application can be easily combined with existing documents.

An effective computer application is one which is welcomed by management because of the benefits it provides to the organization, and welcomed by end-users because it meets all their requirements. Development procedures should ensure that these benefits are provided in the operational system as a matter of course. This applies whether the system is developed from scratch, or based on the use of an application package, or based on the use of functional software, or if productivity aids are used merely to enhance existing procedures.

Development methods are still based on the assumption that the major barriers to developing an effective system are technical; the difficulties of writing thousands of lines of program code to fit seamlessly together in a working application. The emphasis has shifted over the last decade to stress the key role of data structures; the main development problem is now perceived as the analysis and design of complex data structures to fit seamlessly together.

This is still a major requirement and is partly why structured development is necessary, but there is a parallel requirement. During the development process there is a development and fleshing out of the requirements which were originally laid down as the framework for an effective system. They may or may not change but they have to be defined in more and more detail. It is sometimes possible to establish detailed requirements with some forms of prototyping, but prototyping is not always possible.

The expansion of the total application requirements should be systematically tackled before the project moves to the design of the computer system. If there are unwelcome side-effects to using computers, they should ideally be found so that the computer system can be designed to remove them.

A common example in the early days of computing was eliminating the customer ledger card without making any provision for answering telephone queries from the customer. A similar example is eliminating the 'rough book' used in telephone ordering without providing anything else to carry jotted notes about the order. Streamlining a tedious operation can often increase the tedium as more boring work is handled in a shorter time by fewer people. This is a common cause of user grumbles about computer applications.

The design aspects of user-oriented development belong in a separate book but some design procedures are mentioned so that the link between analysis and design can be shown.

The first chapter is an outline of some characteristics of business systems to be considered during development. Many of them are very obvious, some of them to the point of banality. They are written down because they do not seem to be written anywhere else, and accordingly are often overlooked.

Chapter 1

The Nature of Business Applications

INTERDEPENDENCE OF SYSTEMS

There is no such thing as an independent and free-standing computer application; every application in business communicates with other applications either by sending messages and information to them or conversely by receiving messages and information. The mechanized system for processing an order-book will need eventually to pass messages either to a warehouse management system or to a production planning system. It will also need to pass messages to some form of management information system, and when the order has been processed it will need to pass messages to an invoicing and customer accounting system (see Fig. 1.1).

In turn, each of these associated systems will need to pass messages to further systems, to production control or inventory control systems, to financial systems, or to management information systems. These further systems again pass or receive messages to and from other systems in purchasing, production planning, engineering, etc. The administrative and management systems in an organization together form an overall macrosystem which may be tightly or loosely integrated according to business culture and management style, but must always be integrated to some degree.

The conclusion of this consideration is that although the formal terms of reference for developing an application may be 'Produce a mechanized system for the management and analysis of the order-book', the real terms of reference should be 'Produce a mechanized system for the order-book which can be integrated automatically with any other computer systems which present or future business imperatives will demand'.

All business administration and all business information is interconnected. Accordingly all business applications on the computer need to have designed into them 'import-export facilities' for messages to and from other applications.

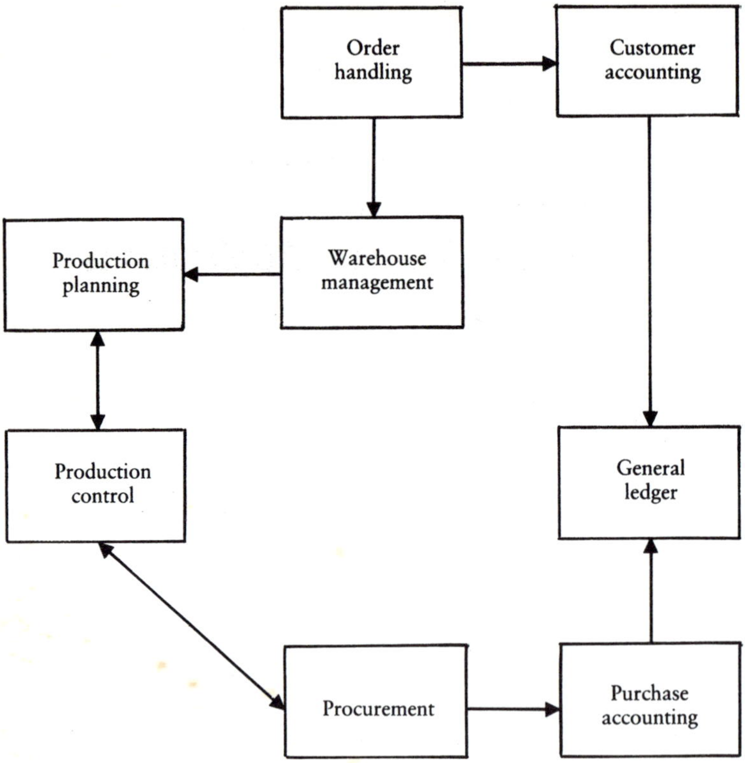

Fig. 1.1 Interconnected applications.

DATA PROCESSING SYSTEMS AND MANAGEMENT INFORMATION SYSTEMS

Most texts on business computing talk of the 'systems triangle' which shows that information from one system is refined and processed to produce further information for a higher level information system (see Fig. 1.2). The triangle usually shows systems at the operational, middle-management, and top-management levels.

This is correct as far as it goes but fails to point out that applications at the operational level are different to applications at the other two levels. A distinguishing characteristic of operational level or data processing (DP) systems is that the proposed system has to be closely meshed into a set of complicated, and usually rigid, manual operations. This applies to all the traditional DP applications from payroll through ledgers, invoicing, purchasing stock control, etc.

Generally the DP system will change the content and number of jobs in the affected office, and it will change the organization and structure of the related manual activities. For instance a by-product of the mechanization of the purchase ledger probably will be that posting figures for the general ledger will be made available some days earlier than with the manual system. This may mean rescheduling the month-end activities of the general ledger section.

Or perhaps a decision to produce invoices and statements in one batch at the end of the working day (so as to sort them by postcode to take the postage discount) will cause problems in the post room, especially if the last post is collected before the end of the working day. If each DP development project does not discover two or three similar problems, it is fairly certain that the business analysis has been too superficial. It has failed to take into account the impact of the computer system on manual day-to-day operations.

MAN–MACHINE BOUNDARIES IN DP APPLICATIONS

As a result of the analysis which is carried out during the design of the system there will usually be various alternatives in a handful of areas for the positioning of the boundaries of the computer part of the system. This positioning of boundaries should be analysed and judged systematically.

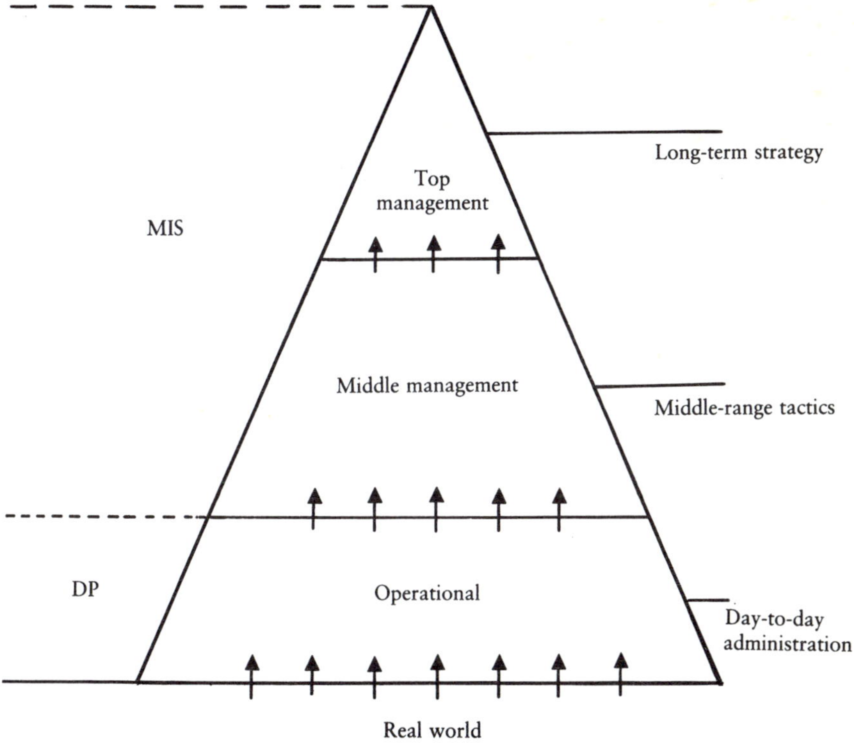

Fig. 1.2 The management triangle.

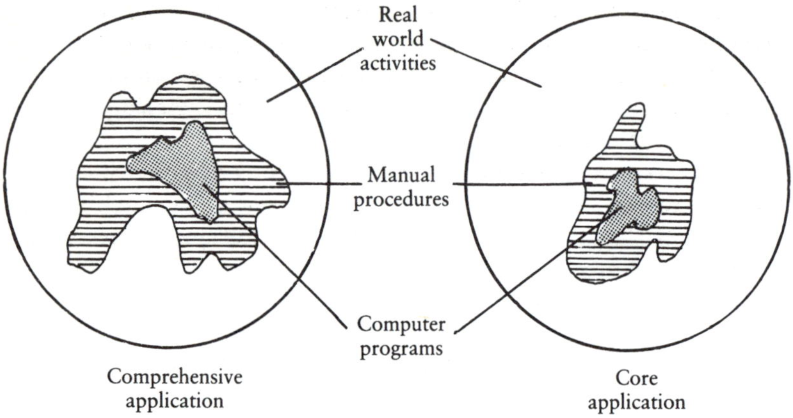

Fig. 1.3 Fixing application boundaries: two different approaches to mechanizing an order book.

For instance, in the mechanization of the order-book one activity might be the allocation of codes to new customers. The usual options are:

1. It may be left as it is in the manual part of the system.
2. It may be handled by the computer system.
3. It may be handled by the credit controller in a different department via a manual request.
4. Via a request generated by the computer system.

Both the efficiency and the effectiveness of the final system depend more than anything on the four or five similar considerations that invariably arise during the design of most applications. They are unlikely to be appreciated by managers, but they may be crucial for users. Fig. 1.3 shows two different approaches to mechanizing an order-book.

The positioning of the man–machine interface for a DP system has more effect on its utility and economics than any other single factor. This seems self-evident, as it determines both the operational costs of the application and the effectiveness of the join between computer processing and manual operations. However it is rarely considered systematically, and there is no check in traditional development methodologies that alternatives have been considered.

SCREEN DESIGN FOR DP

The design of screens has to meet different criteria for DP systems and for the various kinds of management information systems (MIS). For DP the screens can usually contain more information than is considered good practice in

MIS, but above all the DP level is where raw information enters the macrosystem and validation of the information must be very tight.

It should be made impossible for significantly incorrect information to enter a DP system; there are design techniques to check the correctness of most information and the feasibility of all information. They include check digits, hash totals, control totals, range checks, feasibility checks, duplicate entry, use of redundant data, display of associated information, table checks, etc. These design techniques should be used systematically in the design of all DP systems.

TASK MISMATCH IN DP SYSTEMS

The interface, i.e. the screens and the written reports, of the DP computer system have to be closely matched to the mechanics of the manual operation. A well-known case where this was missed was in designing screens for checking-in at airports. The information on the screen was not presented in the sequence in which it was shown on the air ticket.

As a result the booking clerks sent an incomplete screen to the computer whenever the screen sequence did not match the ticket. The system then required them to recall the partially completed screen and to fill in a few more blank fields before again sending it off to the computer. The result was an increase in the time taken to confirm the booking, and a computer system which collapsed under the weight of the 'error' messages it was receiving.

The task on the computer must be closely matched to the task which is presented from the manual part of the system. 'Task mismatch' is a common complaint and is avoidable. Usually the reason is that the programmer never checks with the user the consequences of using a particular style of screen dialogue and cursor management.

BY-PRODUCT DATA ENTRY

Whenever data is fed into a computer with the sole purpose of providing the computer system with information, it is likely to be incorrect. Even with a well-managed and motivated work-force, experience shows that there will be an error rate in the codification and entry of most documents.

On the order-book example, the entry of the order into the computer should result in the production either of a picking list for the warehouse or the raising of a make-order on the shop floor. If there is something wrong with them the manual system will probably pick it up, e.g. the address is incorrect or the customer name is wrong.

If however the information is entered into the computer from a manually produced picking list or make-order, it will possibly be entered incorrectly and will not be spotted until much later. The secret is to use the style of the supermarket check-out and obtain the computer input from some sort of keyboard or sensor operation (e.g. weighbridges, badge readers) that is an intrinsic part of the administration itself.

The default option on all DP applications is that data entry should be based on information which arises as a by-product of some other necessary operation. It follows that ideally there should be no direct data entry of information to any MIS. An MIS should ideally obtain its information exclusively from DP systems. There are few people who have experienced GIGO (garbage in, garbage out) who would quarrel with this point, but there are even fewer who take it into account when developing applications. Managers are especially reluctant to accept it in practice.

LOGISTIC AND FINANCIAL SYSTEMS

DP systems are concerned with either the flow of money through the organization, or with the flow of things, and the designs of the two classes of system are different.

In most cases the prime requirement of a logistic system is the provision of speedy information at the expense of everything else including incomplete input and minor errors. A production schedule must be available for the start of the working day or the factory cannot start. This makes it essential to use design techniques to ensure a speedy flow of data, and to use fail-safe routines to deal with data that is not made available in time.

Financial systems conversely emphasize the provision of correct and complete information, and use procedures to highlight missing information. More important even than correctness is consistency; the information contained on one message must match allied information contained on other messages; a payment must be matched to an invoice, a debit must balance a credit.

The system should be actually designed to be inoperable on inconsistent information. If a payment does not match an invoice, there must be a balancing item either generated by the first stage of the system or provided to the first stage of the system, before the second stage becomes operable.

Logistic systems must be timely and financial systems must be correct and consistent. It is desirable but not essential that the converse also applies, though this is a second priority.

A second difference between the two kinds of system lies in the end-users. Very broadly, the end-users of financial systems are used to dealing with paperwork and with figures while many logistic end-users regard paperwork as a necessary evil.

Logistic paperwork has to be simple and robust enough to cope with the less than ideal environments in which it will be processed. It is good design practice to use prenumbered documentation for the payment of hire-purchase instalments, but it may be bad design practice to use it in a foundry where it is likely to be lost or destroyed.

THE SYSTEMS DESIGN PROCESS

All systems development goes through the following four stages:

1. Defining the application in business and end-user terms.
2. Defining the application in computer terms.
3. Building the application.
4. Installing the application.

The only difference between applications is that with larger and more technically complex applications, the four stages have to be split into sub-stages until manageable units of work are obtained which can be released to the members of the systems development team. On small applications it may be that all the activities and problems in a stage can easily be considered simultaneously so that there is no need for a great deal of development documentation. It might be sufficient to mark the passing of each stage by a short note. An aim in all development projects should be to produce the minimum amount of development documentation that is consistent with control.

In larger applications each stage may be split into dozens of individual items to be considered and activities to be undertaken, and in these cases there is a need for some sort of co-ordinating body. This invariably leads to further documentation, and the entire operation becomes very bureaucratic. It is pointless to criticize the DP bureaucracy involved in developing large systems; there is no alternative if control is to be exercised.

It may however be possible to circumvent it. A system which is twice as complex as an alternative probably takes about four times as much effort to develop because of the overhead costs of co-ordination. There seems to be a kink in the cost curve at respectively one, three, and seven man-years development effort.

The secret in very large projects is to see if they can be split into separate but connected projects, each of which can be managed independently of the others. This point is rarely made in books on systems design.

On large projects an essential step in the development process is to see if they can be divided into a number of smaller projects to cut development costs and to ease project control. This applies whether third generation or fourth generation development tools are to be used.

OVERLAPPING OF PROJECT STAGES

One factor however applies regardless of the size of the project: the four major stages should be self-contained and never overlap (see Fig. 1.4). If the project has been split into smaller independent projects, it is possible to be working say on stage two of the first miniproject while being involved in stage one activities on the second miniproject. This overlap is a good reason to split projects wherever possible as development is less costly and takes less time.

At first glance this banning of overlap between stages may seem to be at variance with what has been written elsewhere in this book; in fact it is not. Within each major stage of development the overlapping of activities is highly desirable, as it leads to both a shorter development time and a more flexible use of development team members.

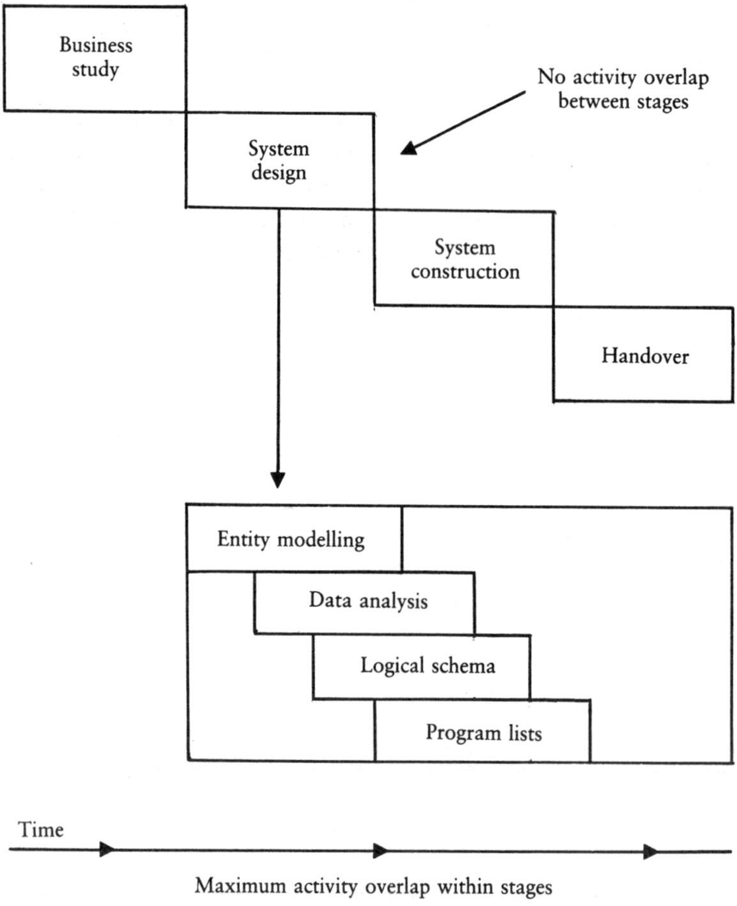

Fig. 1.4 Activity overlap.

In stage three (the building of the application) it is desirable to overlap the writing of one set of programs with the logical testing of another set of programs, and perhaps even with the testing of a series of modules in another part of the system. In stage four, it is nearly always preferable to phase the introduction of the system so that the end-users are asked only to cope with innovation at a rate that they and the organization can accept.

However, there should be no overlap between stages. The business definition of the system should be frozen before the computer-oriented definition of the system is initiated. Similarly this should be finalized before the construction of the system is begun.

This absolute division between the major stages is a main plank in shifting the emphasis of development over to the consideration of management and end-user requirements.

If all these points about the nature of business computing and the design process for applications are considered systematically during the design process, most user criticisms of the final product will disappear.

Technology is changing very quickly and the range of applications is growing even more quickly, so that it is impossible to foresee all the implications, especially for the DP or administrative support applications discussed earlier. However, it is possible to play the percentages so that any eventual faults in the application will be because the developer has guessed wrongly about how best to meet requirements rather than because he was not even aware of the seeds of likely shortcomings.

Chapter 2

Fixing Objectives and Justifying the Project

COST-DISPLACEMENT AND ADDED VALUE JUSTIFICATION

Developing and installing a business application represents a capital investment which has to be justified. Scarce technical skills from the development team and substantial amounts of management time may be expended in the development of a business application, as well as a substantial amount of money. The investment of time and effort and scarce resources can only be justified if the application produces a clear and unambiguous pay-off.

Sometimes the pay-off can be measured in savings in hard cash, perhaps a reduction in staff numbers or a reduction in working capital. Money and skills are expended with the objective of making future reductions in operational costs. The term for this kind of justification is 'cost-displacement'.

Another kind of justification is the generation of additional wealth and profits. For instance, it may be that potential sales of cars are lost because the dealer is unable to locate quickly a car with the precise specification for which he has a customer waiting. If the customer is not able to get the colour combination and the accessories and extras which he requires, he is quite likely to find what he wants in the showroom of a rival marque.

Using a system based on local terminals and a public network wire service, dealers could locate any car which matches or nearly matches the requirements of their customer, even if they are in Bristol and the car is in a Newcastle showroom. Such a system is in use for Rover dealers in the UK.

There is a pay-off to installing this application but it does not rest on cost-displacement, in fact the new system might be more expensive to operate than the system which it replaces. The pay-off is due to the additional business which the dealer is able to process and the usual name for this kind of economic justification is 'added-value'.

In passing, a set of added value benefits is often missed because management do not know enough about computers to realize that they are possible.

These benefits arise when the computer is used to produce information that cannot be produced by other means. Some simple examples are: the use of cumulative sum techniques to spot a deviation from plan well before it becomes apparent to the naked eye or to simple analysis of variance; a statistical analysis of bought and purchase ledgers to help in short-term liquidity forecasting; the use of linear scoring models in trade-off situations. It is part of the job of the development team to make management aware of these possibilities.

TANGIBLE AND INTANGIBLE BENEFITS

The justification of DP and MIS applications is usually more complicated than simply totalling the estimates of various cost-displacements and added values that will ensue as a direct result of the application. The end-result of all business applications should be either cost-displacement or added value, but often this end-result is only an indirect consequence of an application. Figure 2.1 shows the four categories of pay-off.

This is very often the case in MIS applications where the added value of providing management with better and more timely information will eventually be increased profitability for the entire business. It is however very difficult to relate consequential increased profitability to the application which enabled it to be taken.

This kind of indirect benefit is usually referred to in the literature of business computing as in 'intangible benefit'. There is however nothing intangible about it; it can be measured in money terms, or market share, or level of service, or speed of service, etc. In fact, if it cannot be measured it does not exist except as a piece of wishful thinking.

Indirect benefits help to create a situation where a higher level of profits can be taken but cannot be ascribed to a single set of actions or to a single application. Most of the benefits, for instance, of producing sales statistics and market analyses fall into this category.

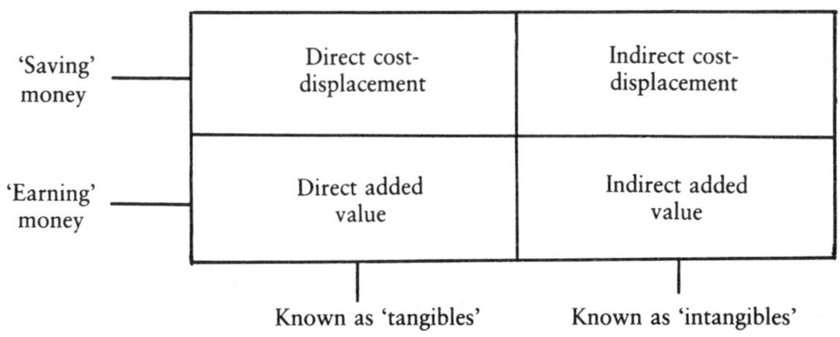

Fig. 2.1 The four categories of pay-off.

Another way of regarding indirect or 'intangible' benefits is as a series of enabling factors, that is removing obstacles to either cost-displacement or added value opportunities elsewhere in the organization. The entirely valid justification for an expensive application may be the provision of operational tools or relevant information to management to enable it to reduce operational costs or to increase margins.

THE NEED FOR COST-DISPLACEMENT OR ADDED VALUE JUSTIFICATIONS

Every application needs a quantified justification to provide a target and a framework for the development process. This is partly because it is sound business practice, partly because it concentrates the minds of the development team to have concrete objectives. It is one of the paradoxes of business computing that an environment where no expense is spared to produce an application of technical excellence invariably produces systems which are technically rather poor. Thomas Watson knew this when he chose 'Think' as the motto of IBM.

If there is no quantified justification for the project or if the development budget is too generous, the project will lose its way as the objectives of the application become wider, different, more ambitious, and eventually self-contradictory. The development process must have a goal. The request 'Develop a system to manage and control the order book' contains the seeds of major problems which will sprout when different sponsors of the system realize that they are seeking different kinds of benefits.

The example known to most DP managers is what invariably happens during the definition of the algorithm for controlling stock levels in an inventory system. The accountant wants minimum capital employed; the marketing man wants maximum customer service; the distribution manager wants something that will fit with existing racks and bins; and the production manager wants lengthy production runs. Most of these requirements are not compatible.

In the case of the order-book, incompatible objectives might be as follows:

1. Staff reductions.
2. Handling an increased order volume.
3. Processing orders more quickly.
4. Improving liaison with the warehouse.
5. Provision of statistics and analyses of orders.
6. A mix of the above.

Some of these might prove to be incompatible when the analyst comes to design the overall computer system.

A more meaningful request to mechanize the order-book is: 'Develop a system to do the following:

1. Reduce operational costs by at least 20 per cent.
2. Process double the volume of orders with no increase in staff.
3. Provide a trend analysis of orders to help in timely procurement.

The system is required by the first quarter of next year, and whatever is expended on development should be recovered in two years.'

Common sense indicates that the system developed according to the second set of requirements is more likely to be a commercial success, and common sense would be right. But experience shows that the second set of requirements will also result in an application which is technically more elegant and robust.

It is the job of management to be unambiguous in their request. If they are not the application will need luck to succeed.

SOME COMMON MYTHS ABOUT ECONOMIC JUSTIFICATION

'An application is either of the cost-displacement or the added value type.' This myth is usually implicit in loaded questions such as 'Do you want the computer to save money or to earn money?' The short answer should be 'both'. The cost-displacement type of application, when properly designed, invariably brings with it added value benefits. This is because it is difficult to increase efficiency (cost-displacement) without also increasing effectiveness (added-value).

'Intangible or indirect pay-offs have more potential than tangible or direct pay-offs, because the intangibles spread through the entire organization whereas the direct benefits apply only a small part.' This myth is often implicit in the waffle of the novice business analyst, e.g. 'If the system only increases sales by a small percentage the resulting profits will swamp the savings associated with the salaries of a couple of clerks'.

Apart from the general unreliability of statements based on unprovable 'ifs' (if pigs had wings . . . etc.), the rest of the statement is also likely to be incorrect.

Cost-displacement savings should always provide a base from which other cost-displacement savings may be obtained. A cost-displacement saving is not a platform but a rung on a ladder which permits movement to the next higher rung. Once the savings have been made a new situation develops and further savings invariably become possible, sometimes referred to by DP managers as 'creeping mechanization'. After two or three rungs, a virtuous spiral of productivity sets in for both labour and capital.

Additional cost savings arrive automatically as the system beds into the

Table 2.1 An example of a wish-list

Objective	Ranking	Weight	Comment
Reduce staff	3	10	Only 1 likely saving
Reduce inventory	4	10	Low cost at present
Reduce operation costs	5	10	Not a problem
More capacity	1	30	The market is booming
Remove bottleneck	2	30	
Faster processing	7	5	
Better service	6	5	
	Total	100	

office culture, and as its potential is seen by operational staff and management. There is a learning curve for the organization as well as for the individual end-users. Additional savings of between 25–50 per cent of the original savings often accrue automatically over the initial two years of operation.

The really important additional savings, however, arise from the gradual extension of the original system to incorporate more activities and functions. The application should therefore be designed so that this kind of extension is easy. It spreads the efficiency of the original application in a benign way through the organization.

This 'contagion' phase of DP and MIS is better known in office automation applications but it applies to all IT applications. It should be catered for in the initial design.

One should always obtain agreement on the business objectives of the application before the project starts and one method is to ask each person involved in setting the objectives to complete a 'wish-list' (see Table 2.1) with the help of a team member. Any misconceptions about the project will then emerge when the wish-lists are compared.

Every manager whose role is affected by the information to be processed should be approached, and not just the manager with direct responsibility. In the case of the order book it might be advisable to talk to the accountant, the marketing manager, and the distribution manager, as well as the commercial manager who is responsible for the order book. In the last resort the line manager has to decide, but the development team should present him with a global rather than a departmental picture.

In the example, a well-administered order-office is faced with a surge in the volume of orders and is losing business to the competition. Thus, while more efficiency is welcome, the main requirement is perceived by the author of the list as the provision of more administrative capacity so that additional business may be obtained. The main justification is direct added-value, with direct cost-displacement seen as being roughly half as important.

This is important as it places an emphasis on speed of processing so that twice as many orders may be handled by the same number of staff. If the emphasis had been on eliminating staff positions, the emphasis would have been on mechanizing as many fringe activities as possible. The two systems are quite likely to be different in design and operation. The split between batch and transaction processing might be different, with the emphasis on speed demanding the maximum functionality in transaction processing while a requirement to reduce staff could lead say to the batch production of some documentation overnight.

This point of orienting application architecture to business objectives is rather glossed over in traditional development methodologies.

Wish-lists contain different kinds of objectives for every application because every application is unique. It is the job of the development team members to ask leading questions to enable the different objectives to be entered in each individual wish-list. A few examples of possible requirements are given below for the four categories of justification. These should be used to initiate the interview with management.

Direct cost-displacement: less staff; less inventory; less consumables; reduction of overheads; avoid future staff hiring; replace scarce skills; reduce overtime, reduce WIP.

Direct added-value: increase capacity; de-bottleneck some procedures; speed-up processing; raise level of service; maximize profit for seasonal or fashion products; introduce new procedures (e.g. anti-pilfering).

Indirect cost-displacement (enabling factors for): least-cost deliveries; more reliable information; information retrieval for marketing; less-expensive procurement.

Indirect added value (enabling factors for): more up-to-date information; more detailed information; *ad hoc* information service; professional productivity; management productivity.

QUANTIFYING THE PAY-OFFS

The justifications for a DP or MIS application should be quantified, partly as previously stated to provide a framework for the development of an effective and technically robust application, partly to provide a yardstick of the value of the application.

Quantifying benefits in money terms also automatically fixes priorities for competing applications. Some organizations employ complicated systems of ranking applications, when all that is needed is a policy of mechanizing first the project with the biggest pay-off.

Provided that an approximate quantification of the benefits gives a project target and yardstick, any additional precision is meaningless. If it is obviously possible to halve the cost of producing an invoice by using a new computerized system, it is spurious professionalism and bad management to stall until the organization and methods (O&M) people have produced an analysis which shows that the cost probably will be reduced by 55 per cent.

The future is so imprecise that their guess is very likely no better than any other. In addition, as the project develops it is likely that further benefits will emerge, and if the team is restricted by over-precise statements about the system objectives, it will not take them. This is a common fault in large mainframe installations.

The best way to justify direct savings is to ensure that the minimum savings are adequate to meet the return on investment, or payback period, that is set by management. The payback period should not be less than two years (or the team cut too many corners) but equally it should not be very much more.

This means that if the total cost of installing a new system is £50 000 the system should bring £20 000–£25 000 p.a. in direct benefits – if it is justified exclusively on direct benefits. Once the development team can see benefits of £20 000 p.a. and management agrees with them, the project should be given the go-ahead. There is always a better application but life is always too short to find it.

As far as direct or tangible benefits are concerned the estimates of the likely pay-off should be quite accurate. Everyone knows what it costs to employ a clerk, and everyone knows the daily cost of keeping a case of inventory in the warehouse.

Added-value direct benefits are almost as easy to ascertain. If the order office is processing only 90 per cent of the orders and delivery is stretching from weeks to months, there will be lost sales as clients move to alternative suppliers. The consequential lost sales can be ascertained from the changing ratio of order-enquiries to firm orders.

The conversion of recovered lost business into an added-value benefit is part of the work of the development team, but (and this is an important point) it is not their job to assume that new business can be gained by a better system. That judgement is the prerogative of the line managers responsible for obtaining new business.

One of the common dangers in systems development is the tendency of team members to assume the duties of management in evaluating and introducing innovation. This cannot be cured but its effects can be minimized by an absolute insistance on both management's and users' participation in the development process.

Even more common is the tendency of management to become involved in strategic technicalities (e.g. choice of supplier, siting of machines in a network) because of the sums of money that may be involved. Neither can this be cured, but the inevitable disaster can be avoided by giving the DP manager the right of veto on all proposed purchases.

The justification of a system on its indirect (so-called intangible) benefits is difficult. Firstly, although the benefits can be quantified, the reliability of the estimate is lower than with direct pay-offs. A foolproof system of estimating delivery dates will obviously bring in additional orders, but the estimates of the additional business might differ by a factor of four or five to one. No one really knows until it happens.

Secondly, the estimates of increased business may be based on a variety of procedures only one of which is provided by the order-book system. Judgements, therefore, have to be made first about the amount of increased business, and second about the effect of mechanizing or not mechanizing the order-book.

In other words there is scope for argument and different opinions whenever indirect benefits are quantified. The only possible way to proceed is to show that there are such overwhelming benefits that no matter what interpretation is placed on the figures there is always a case for mechanization.

The installation of CAD in automobile design offices means that the design engineer is able to consider twenty or thirty times as many options as was possible with the old manual system. This eventually and inevitably will reflect in the final product. Whether the product is 'better' by 10 or 20 per cent is impossible to say, but it is possible to say that without CAD it would not be possible to produce competitive designs.

Unless there is an overwhelming case for indirect savings the organization is well advised to ignore them. Certainly they should not be given the same weight as direct pay-offs, even if a consensus exists that the indirect benefits will accrue. If there is no consensus, there is probably no basis to proceed as the pay-off is too tenuous. This means that if a pay-back of two years is required for projects with direct pay-offs then a pay-back of a few months is expected from projects based on indirect benefits.

When one talks of pay-backs of this magnitude one is effectively saying that indirect benefits have to be so self-evident and massive that the case for mechanization hardly needs arguing once the application has been defined.

For evaluating indirect benefits the procedure is to obtain a consensus on what is going to be the effect of the application on management effectiveness and/or professional productivity. The only people who can evaluate managerial effectiveness are the managers themselves, so an application based on the indirect benefits of a management support system is effectively accepted or vetoed on management opinion.

For professional workers, the procedure is to define a 'work product' (a plan, a drawing, a report, a set of market analyses). It is then a simple task to establish the cost of producing a work product under the present system and the estimated cost under the proposed system.

Usually one finds that professional staff are not reduced but that the same staff consider more options in order to increase effectiveness. One should attempt to turn the increased productivity into commercial terms. One cannot say that if fifty alternatives can be considered instead of five that the benefit is equal to ten times the current cost of the five. A geometric progression gives a first rough and ready answer which is often the best that can be obtained.

Increasing productivity by four probably doubles the value of the professional work-product, increasing productivity by nine or ten triples the value of the work-product, and so on. The questions about added value are often as meaningless as the question 'how long is a piece of string?', but they have to be answered and the above technique is a way of replying 'long (or not long) enough for the job in hand'.

Chapter 3

The Role of Management in Business Analysis

Once the objectives of the system have been determined on the wish-lists, the development team and management have to fix in general terms the range of the application, and subsequently to define the outlines of the information requirements for both DP and MIS applications. More detailed definitions of the application boundaries can be determined later in conjunction with the end-users, and the details of management information requirements can also be fleshed out at a later stage.

Unfortunately, the computer professional does not distinguish adequately between the total application, which is a mix of manual and mechanized procedures, and the collection of computer programs at its centre. Most hardware suppliers and many consultants have never grasped the difference between a set of working programs and a working application, probably because they have never actually met the users of a poor system with excellent programs.

Every application is made up of many related procedures and there is a double decision to be made about each of them.

1. Is the activity to be included in the application?
2. Is the activity to be mechanized?

Taking the order book as an example (see Fig. 1.3), the list of core and fringe activities would number several dozen and might include: order enquiry, order acceptance, accepting a new customer, checking the credit of existing customers, modifying information about customers, etc. All of these activities might be included in the application, or many of them might be excluded. If they are included they have to be woven into the fabric of the application even if they are not mechanized.

FIXING THE BOUNDARIES OF THE APPLICATION

The initial management activity in defining the application is to fix the boundaries of the total application. The subsequent fixing of the boundaries of the computer kernel of the application should be carried out later by members of the development team and end-users. This is treated in the following chapter.

Fixing the application boundaries is based on the idea of the 'activity chain' which can be documented by the production of a broad-brush data flow diagram (DFD) if DFDs are already in use. An activity chain is a series of activities each of which can be initiated only when previous activities in the chain have been completed.

The DFD is the sign that the overall boundaries of the entire system have been defined. Like the wish-lists, its production ensures that a task has been systematically performed. If a DFD only tells people what they already know, it can be substituted by a list of the activities in the activity chain. The activity chain list is usually more easily understood by management and it indicates the sequence of activities better than a DFD.

The entire process of systems development should be controlled managerially and technically through the production of deliverable documentation. This means that key development activities should produce documentation summarizing their conclusions. A wish-list is deliverable documentation.

Assume that the ABC company is selling high-value items ex-stock to business customers or retailers who are invoiced at the end of the month. The order-office has to deal with the following principal activities:

- Order enquiry.
- Order received.
- Order acknowledged.
- Order passed to warehouse.
- Order picked or partly picked.
- Optional balance order raised.
- Order dispatched.
- Advice of delivery dispatched.
- Order accepted by customer.

The above list is an activity chain with each activity depending on the successful completion of a previous activity. At the end of the chain a message is sent to the customer accounting office to raise an invoice. Messages are also sent to the stock control system when the order is received and when it is picked so that the book stock records and physical stock records can be adjusted.

There are additional messages which are passed to MIS applications at the end of each day, and there are further warning messages which are passed to operational management for situations such as stock-outs and likely stock-

Table 3.1 Activities covered by order handling system

Section	Activity	Included?
Order office	Order enquiry	√
	Order accepted	√
	Maintain stock records (book)	?
	Accept new customers	?
	Instruct warehouse	√
Warehouse	Produce despatch documents	√
	Stock out notes	√
	Partial deliveries	√
	Maintain stock records (physical)	√
	Deal with returns	√
	Place make orders	?
Order office etc.	Balance orders	√
	Request for billing	√

Note: Completed in any order when management think of activities. Most activities should be listed by the team in advance of management meeting.

X = Exclude
√ = Include
? = Possible inclusion

outs, abnormal orders, broaching credit limits, and differences between book stock and actual physical stock. Table 3.1 shows the activities covered by an order handling system. *Note*: This is a simplified list which ignores ancillary activities for modifying and controlling the major activities. If all the detail were included the text would be unreadable to anyone not familiar with the business. The detail in real-life applications is always greater than it can be in a textbook.

It is a joint management and development team activity to decide the boundaries of the proposed application which, after the evaluation of the wish-lists, has now been defined as the 'mechanization of the order handling procedures primarily to provide capacity to handle double the volume of orders with existing staff'. This is now the mission statement of the development team.

Some of the basic decisions to be taken are as follows:

1. Are warehouse administrative systems to be included or is the system to deal solely with the order office?
2. If warehouse systems are to be included should they be pushed further to cover distribution activities?
3. Should a stock-control system be constructed from the order-book and the stock-ledger applications?
4. What should be the connection to the present customer accounting administration, especially if it is only partially mechanized?

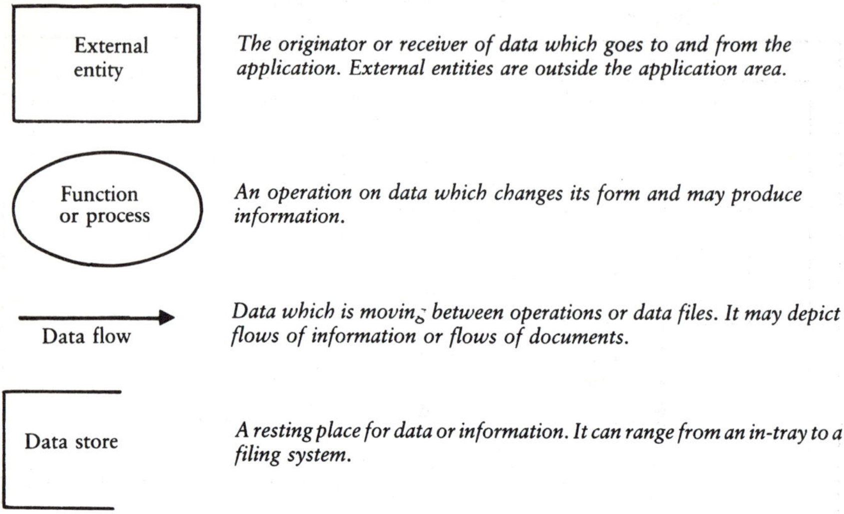

Fig. 3.1 The elements of a data flow diagram.

5. What type of additional information should be made available to operational management for control purposes and crisis management?

The first step in fixing the boundaries of the application is for the development team to list the activities in the activity chain and to decide with management if they are to be included in the initial application. The implications of including or excluding activities have an effect on the design of the mechanized system, which can be explained to management by the development team.

For instance, although the overriding objective in the order-office might be to cope with increased volumes of orders, it is possible that extending the system to incorporate warehouse activities could probably be done for a very slight increase in development costs and would bring substantial increased benefits. Provided that the initial objectives are not jeopardized it would be poor management to turn down the chance of a bigger pay-off.

Often the definition of the activity chain is carried out using a broad-brush data flow diagram (DFD). It is a matter of preference whether the DFD should be a working document in the team discussion with management. It may be that a straight list of activities is easier for management to understand.

There are only four symbols in a DFD. These are shown in Fig. 3.1.

These can be used to depict the broad-brush definition of the application as shown in Fig. 3.2.

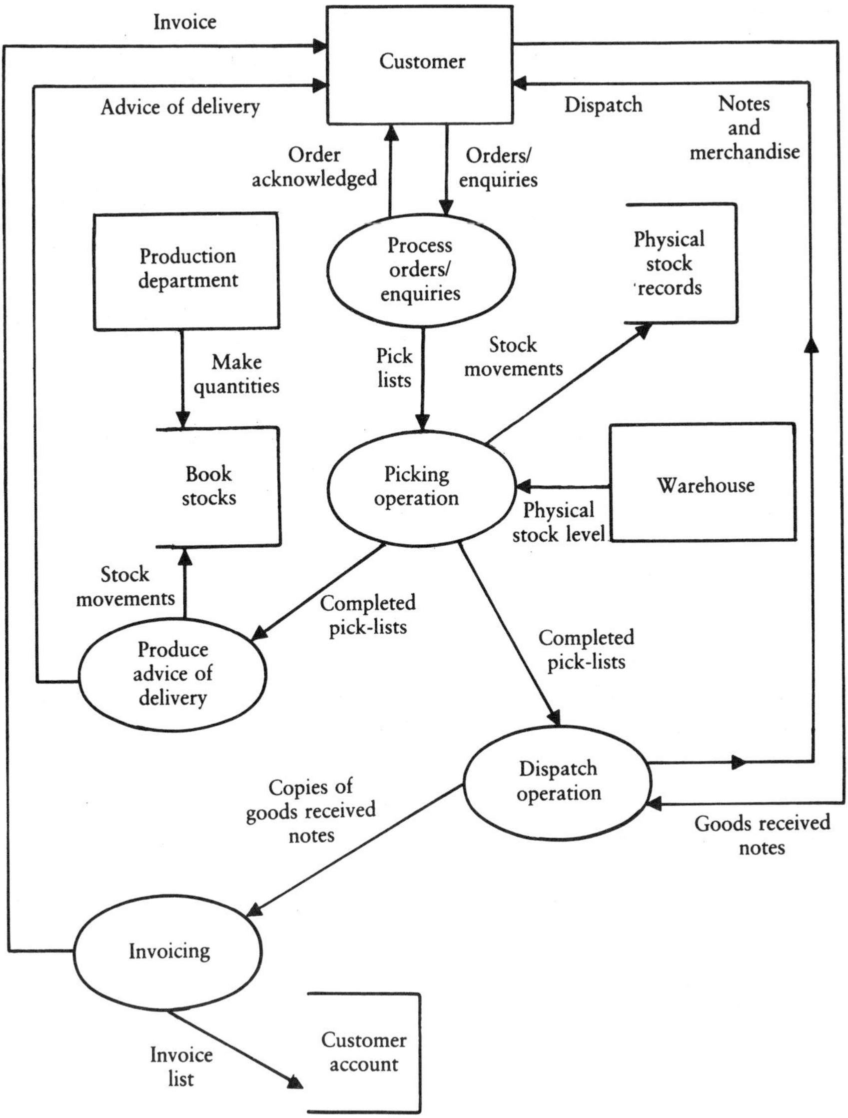

Fig. 3.2 Top level data flow diagram.

DFDs operate at many levels of detail. Obviously only the major procedures are shown in the top level DFD, but it is still a moderately 'messy' document. A minor variation with highlighted major documents is shown in Fig. 3.3.

It is possible to use an even broader diagram called a context diagram. It usually only tells managers and users what they know already and serves to

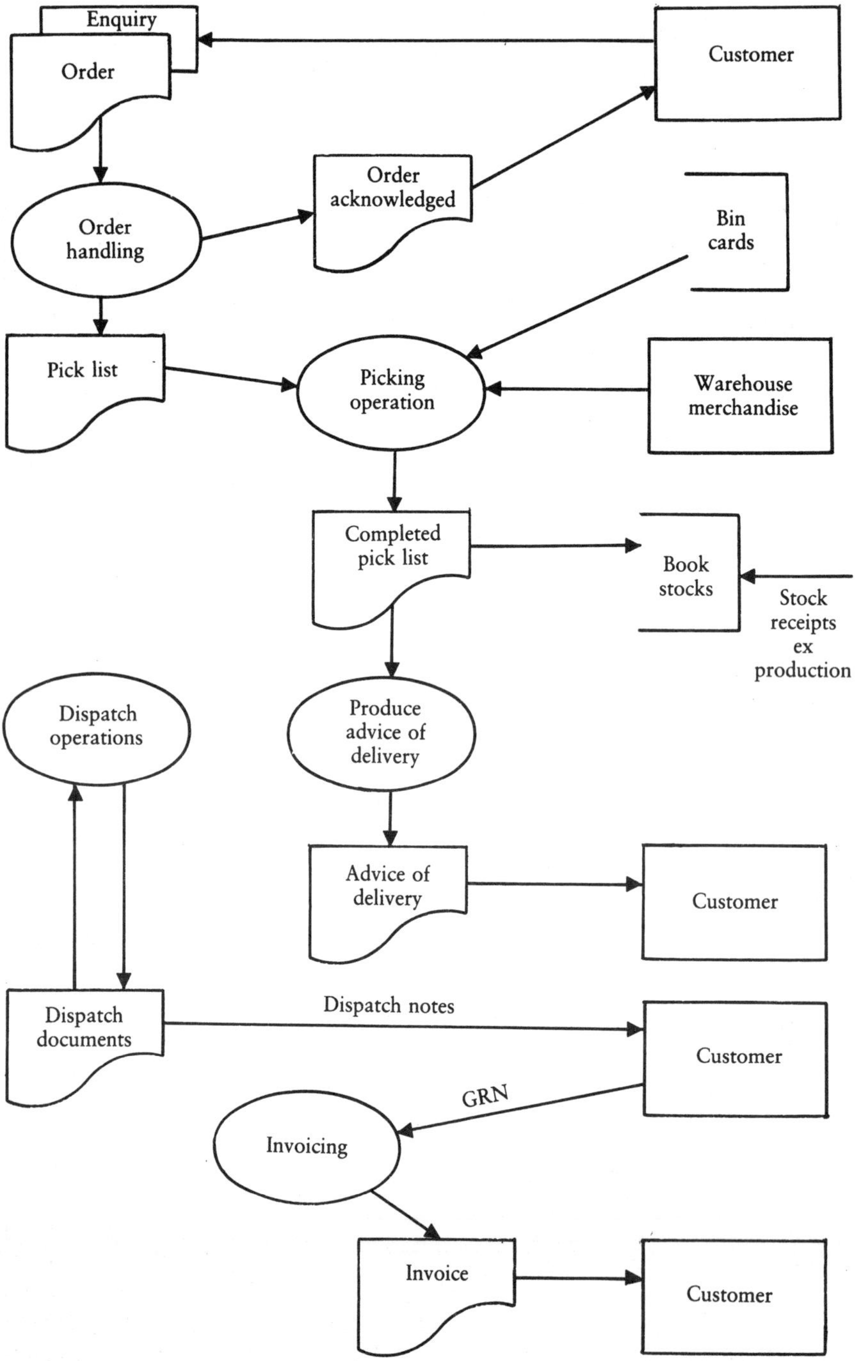

Fig. 3.3 Alternative form of DFD.

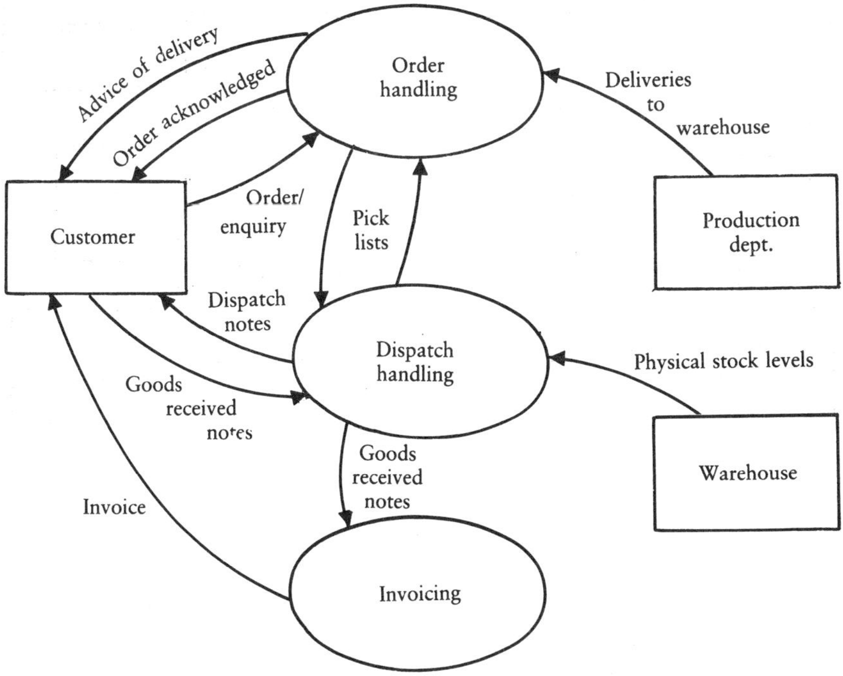

Fig. 3.4 A context diagram.

educate the team rather than management or end-users. But it is still useful to document the decisions taken on application boundaries. A context diagram is shown in Fig. 3.4.

The choice of deliverable documents to show that application boundaries have been fixed are therefore as follows:

1. List of activities in the chain.
2. Top-level DFD.
3. Modified DFD.
4. Context diagram.

None of them is better than the others, and the one ultimately chosen should be the one that takes up least management time.

Management has to be guided by the DP experts in fixing the application boundaries. It should be noted that it is not the computer system which is being defined but the overall application, which is a mixture of computerized and manual operations. As the operating and development costs depend largely on what activities are incorporated into the computerized sytsem, it is perfectly safe for the manager to indicate the overall application boundaries without thinking about costs. He is effectively defining an application area within which the team will need to develop a computer application. In short he is doing his job.

DEFINING INFORMATION REQUIREMENTS

The first stage of application definition is now complete as far as management is concerned. However, there is another task for user management in the first stages of the business study, namely to state what are the eventual information requirements from the system itself and from the MIS which will be constructed on its back.

As already stated there is no such thing as an independent system, and it is highly desirable that MIS applications should obtain their basic information from related DP applications for reasons of cost and reliability. This is the stage where the foundations should be laid for the eventual integration of applications. Many analysts will accept this assertion as a piece of desirable theory which is unfortunately almost impossible to put into practice, because end-users and especially managers are quite unable to define so early in the proceedings what their eventual information requirements will be. This is only because they are not asked the right questions.

The eventual integration of applications should be tackled right from the start of all application design, i.e. as soon as the development process reaches the stage of ascertaining requirements of DP applications and the associated MIS applications. This is done by ascertaining requirements in terms that the manager understands.

All business IT applications should be a model of the real world which is inhabited by real people and real institutions engaged in real activities. The real world is inhabited by real people, placing orders for real products, and taking delivery of them on real lorries, and paying for them with real money. The messages which flow through the order-book application reflect these real-world people and real-world objects and their actions. The job of the systems designer is to build a model of the real world so as 'to hold the mirror up to nature', and if he does not his application will malfunction.

This has been recognized by DP theory since Chen's papers in the mid-1960s, and a large part of the theory of systems development is concerned with the concept and techniques of building models of the real world. It aims to produce an information model which faithfully reflects people, objects, institutions and activities in the real world. It does this by listing all the real-world 'entities' involved in an application, and by listing their 'attributes' or information about them, which should be coded into the DP application.

Thus 'John Smith' is a real-world entity with arms and legs and a house and other attributes, which for the order-book application might include address, customer code, credit limit, type of customer, telephone number, contract details, etc.

…odelling approach is based on listing the entities in an application …ining the information which is associated with them. Unfortunately … approach fails to distinguish between objects and concepts on the one …and and the activities which they combine to carry out on the other. In DP theory, objects, concepts and activities are all defined as entities; an entity is anything about which information is gathered. A DP analyst might list the entities in the order-book application as follows:

1. Customer.
2. Customer order.
3. Customer enquiry.
4. Product item.
5. Warehouse.
6. Production department.
7. Invoice.
8. Picking list.
9. Order acknowledgement.
10. Dispatch documentation.
11. Advice of delivery.

Apart from the fact that there are too many entities, there are weaknesses in this view which become apparent during systems development. The first is that managers generally know all about objects and concepts (e.g. customers' classes of customers, product items and product categories) but they generally do not know about the details of the activities which these objects cause to happen (e.g. ordering, raising balance orders, goods inwards procedures, order enquiries).

It would be surprising if the manager did not know about objects and concepts – he himself generally talks in terms of knowing the market, or the industry, or the trade – that is how he makes his living. It would be even more surprising if he knew about the details of the activities that make the wheels go round, there is simply too much detail for one man to master.

When a manager is asked to define his information requirements as regards the order-book he never (and this really does mean 'never') is able to visualize the sort of information that might be useful to him once the mechanized system has been developed.

Conversely, if he is asked what he needs to know about customers, or products, or points of sale, or warehouse stock, or methods of distribution (or any other objects and concepts) he is able to state comprehensively the kind of information he will require. Much of it will be contained in its raw form in an order, or an order enquiry, or a picking list, or the document recording some other activity, but the manager generally is not concerned with where it is found, he just wants to know what it is.

Once the wish-list and the DFD have been completed the manager can be taken on to the object listings. These, as the name implies, are lists of information which will be required for each object and class of object that has to be modelled in the application.

If there is any difficulty in distinguishing between objects and activities the key determinant is the degree of permanence of the information. In the real world objects are permanent and activities are ephemeral. Customers and products vary only slightly from year to year but individual activities only last for a few days or weeks.

OBJECT LISTINGS

The first class of object is a tangible something in the real world. In an order-book application these are typically:

(a) Customer.
(b) Product.
(c) Order office (point of sale).
(d) Warehouse (stock-point).

None of the activities in the activity chain is considered in ascertaining information requirements, only the objects which carry them out in the real world. This is different to most people's practice in traditional methodologies where information requirements are laid out for every box and line on a DFD.

For each individual or primary object, the first sets of requirements to be ascertained are the information or attributes to be coded into the system (see Table 3.2). These would include name, address, delivery address, customer code, telephone, telex, Fax, etc. and terms of trade. It is generally worthwhile designing a special form on which these details can be entered.

Next, the information which is generated by the system as a by-product of its processing is tackled. It would normally include credit limits, list of orders placed, deliveries to date this year by value, outstanding orders, etc. (see Table 3.3).

Once these have been defined for all the primary objects the next stage is to define the classes of objects about which information might be required. For

Table 3.2 Static and system-generated information

Static for individual objects	Derived for individual objects
Static for classes of objects	Derived for Classes of objects
Information for joint object classes	

Table 3.3 Object listing for primary objects

Primary object systems name	Customer	
Data required	Given	Derived
Name/address	Y	
Code	Y	
Delivery address	Y	
Credit limit		Y
YTD orders		Y
Orders in hand		Y
etc		

customers, this would include class of customer, customer area, industry category, discount category, etc.

Finally there is system-generated information required for each class of objects (see Table 3.4). This might include sales statistics, margins, categories of sales. Note that this last type of information can refer to two different classes of objects at the same time (e.g. sales to DIY stores (customer class) of electrical equipment (product class)). When this type of 'joint object' information is required it should be listed in a summary of likely requirements for *ad hoc* information, and not as an attribute for an object class.

Information gathering should be kept as informal as possible; formality frightens the life out of managers even more than it frightens end-users. The object lists can be completed in a series of meetings so that the manager is

Table 3.4 Object listing for class of object showing the object classes and required information

Object classes of	Customer	
Groupings	Given	Derived
Customer type	Y	
Region	Y	
Industry category	Y	
Major/minor		Y
Payment method		Y
Information needed		
Sales by value		Y
Budgeted sales	Y	
Gross margin		Y
Cost of sales		Y
Analysis of variance		Y
etc.		

reassured that if he forgets something it can be added later – as it can until the end of the business study.

At this stage the first sets of information requirements should be listed into the five categories indicated above, i.e.:

1. Given information for individual objects.
2. Derived information for individual objects.
3. Given information for classes of objects.
4. Derived information for classes of objects.
5. Derived information for combined object classes.

FEASIBILITY STUDIES

One of the irritations in IT is the lack of a standard vocabulary. Many writers use the term feasibility study to mean what this book entitles the business study. Certainly the aims of the two are identical. They are as follows:

1. To ensure that an application is technically feasible.
2. To ensure that it is acceptable to the end-user.
3. To ensure that it meets the requirements of management, and that the costs of development are commensurate with the benefits.

It is of course possible to fulfil these aims with varying degrees of precision and certainty. If a full business study is likely to cost several thousand pounds or even tens of thousands of pounds it may be desirable to carry out a form of *preliminary business study* to find out as quickly as possible if an economic solution is likely to be found.

Not that once the problem has been defined in business terms and operational terms, an experienced DP professional will know if a solution is possible and he can cost approximately the project development. The final costing is carried out at the end of the second stage, (system design), but at the end of the first stage the DP professional is invariably in the position of being able to provide approximate costings for the final system, or alternatively a very precise cost for the second stage. The preliminary business study is also called a feasibility study, and it is sometimes worth incorporating into the business study proper.

The procedures are exactly the same as those outlined in this and the following chapters for the business study proper, but the remit of the team is to get the best possible answer for the expenditure of not very much money. The wish-lists and DFDs and object listings are completed in broad brush terms.

If a preliminary study is included in the project it should vary between using one person for one week to using three persons for three weeks. It should provide roughly the level of detail that is submitted by systems houses when tendering to carry out development work. The time-scales and budgets are usually within 25 per cent of what is obtained from a full business study but can sometimes double or even occasionally triple as more information is uncovered. That is why many systems and software houses work solely on a time and material basis.

The deliverables so far discussed for both full business studies and short feasibility studies are as follows:

1. Wish-lists resulting in a 'mission statement' from management.
2. Broad-brush DFDs or alternatives.
3. Object lists covering the five kinds of information.

Chapter 4
Analysing the Existing System

WHY MANAGERS SHOULD NOT BE INVOLVED

As stated in the previous chapter, it is the responsibility of management to fix the boundaries of the application acting on the advice of the development team. The practical reason is that fixing the range of the application is a business problem where administrative and organizational factors outweigh the technicalities of the proposed application. It might be for instance that business imperatives demand that the application must be developed in the next six months and these imperatives must override the technical view that a much more effective and wider-ranging application could be developed in a year. Or the manager may wish to extend the boundaries of the application to ensure that two sections of a department work together more closely.

But it is unwise to ask management to become involved in the detail of the systems design process. Once the boundaries of the overall application have been fixed the role of management is best confined to approving or rejecting the proposals that are made in the more detailed definitions of the application. This is partly because the scarcest resource in any organization is the time of the senior managers and it should not be spent on work that can be done by other members of staff.

It is also because the detailed design of applications is an activity that managers generally are not very good at performing. They have no knowledge of the details of administration and are unaware of the pitfalls, nor are they aware of their own ignorance. They tend to think that the actual application is no more complex than the sanitized summaries that are used to brief them. They think for example either that there are no nuances in administration, or that the nuances are exactly as they were some years earlier when the manager actually worked in the section.

It is usual in the modern world for procedures to change and the rate of change is increasing. This is partly due to new technology and partly to commercial innovations as market pressure encourages new services. The information base of applications, that is the information processed by application procedures, fortunately changes more slowly than do the procedures themselves, which is one reason why DP development methods emphasize the importance of defining information. The information base of a mechanized warehouse application, for example, is fundamentally the same as that used for manual control, but the procedures of control are very different.

Most procedural changes are evolutionary and gradual so that someone who was familiar with the details of an application five or six years earlier will be unaware that they have changed. Nothing changes on the organization chart and the same messages are still passed to outside departments.

THE END-USER'S ROLE

Once the extent of the application has been fixed, the subsequent fixing of the border between computer processing and manual processing of information is a matter for the eventual user of the system – the only people in most cases who are up to date on the current workings of the application area, and perhaps the only people who can appreciate the implications of including or excluding marginal activities from the mechanized portion of the system.

For instance, the elimination of the 'rough book', on which telephoned order delivery instructions (introduced six months earlier) are jotted prior to the production of order documentation, would be unwise, especially if the rough book is used to deal with queries from the customer about delivery. Similarly, the end-user is likely to be the only person able to indicate how the latest form of special offers should be processed.

It should be emphasized that an application is not merely a set of computer programs operating on a database, it is also the manual activities and corresponding documents which surround the computer system. The smooth running of the eventual system will depend partly on where the boundaries are fixed and partly on how well the computer system is melded with the surrounding manual activities. The effectiveness of controlling potential bad debts might depend on the link between personalized telephone calls and the production of standard reminders and first and second letters on the computer.

HOW TO ANALYSE EXISTING PROCEDURES

In most cases the design of a DP application (as stipulated by wish-lists, DFDs or activity chains, and object listings) cannot be initiated before the existing system, either manual or an obsolete mechanized system, is analysed. This is especially the case if the justification for the application is direct cost-displacement.

The usual method in DP analysis is to produce increasingly detailed DFDs. The weaknesses of this approach are that it provides a barely adequate answer, and it results in masses of documents which are hard to understand for anyone not immersed in the detail of the application.

The analysis of existing procedures should be based on the systematic analysis of existing messages. It should be easy to understand, and it should provide all the information necessary to ensure that the new system will cope with every eventuality that is currently processed.

Message analysis emphasizes the user-view of the application. It has the advantage of being immediately understood by end-users despite the fact that it is more precise and provides more detail than do DFDs. At one time it was widely used, but seems to have been substituted by DFDs which are oriented to developing correct computer program specifications.

Interviews with users to obtain information about procedures are inadequate. Years ago in an invoicing application the staff said that no one could purchase shot-gun ammunition without a firearms certificate, and every certificate represented a customer. The analyst accordingly merged the records on the new system. It later turned out that the USA PX had no certificate, and the CWS had a single certificate which covered sales to all its retail societies. A message analysis would have found the exceptions by noting that there was a special form used in each case.

THE MECHANICS OF ANALYSIS

There are three documents used in the analysis of existing procedures. They are as follows:

1. The message list.
2. The labour balance.
3. The message flow chart.

These documents are used when it is necessary to analyse existing procedures as a basis for designing a new system. In many cases there may be no point in analysing something which is to be replaced, and even where an analysis is required it should only go down to the level of detail necessary to provide specific answers.

The message list

Messages are packets of information that travel from point to point in the application area. For purposes of analysis they fall into the following categories:

Table 4.1 Example of a message list

Line no.	Message name	Type	Volume	Peak
1	Order	O	150/day	90 in a.m. post
2	Order	T	100/day	30/h
3	Order	TX	5/day	
4	Order enquiry	O	15/day	
5	Order enquiry	T	80/day	30/h
6	Balance order	F	8/day	
7	Order acknowledgement	F	250/day	
8	Advice of delivery	F	240/day	
9	Customer query	T	90/day	
10	Customer query	L	5/day	
etc.				

Type: F = Form; L = Letter/memo; T = Telephone; TX = Telex; O = Outside document

1. Information carried on forms.
2. Textual information as carried on memoranda and letters.
3. Verbal information usually arriving via the telephone.

The list contains the titles of all the messages which enter into the application area, or are sent out from the application area, or which originate and terminate within the application area (e.g. working documents) (see Table 4.1).

It saves time in later stages if a copy of all forms is obtained with working information written on them. Specimens of typical memos and letters should also be obtained. If the telephone carries a significant amount of traffic a telephone log should be set up, either by sampling or by asking users to note the time of calls and to categorize them into about five classes (e.g. queries on order status, requests for quotations, etc.).

Most of this additional information should be filed for use in later stages of the development project. The immediate object is to establish a list of all messages which are processed in the application area along with a note of the numbers involved and the timing.

The labour balance

The objective of the labour balance is to establish the allocation of time and effort in performing the various tasks which constitute the work in the section under analysis (see Table 4.2).

Each row in the matrix represents a person employed in the application area and each column represents a message which is listed on the message list. Most administrative and business work has peak periods, and accordingly there should be two labour balances established, one for the period of peak

Table 4.2 Section of off-peak labour balance

Line no.	Position	Name	1	2	3	4	5	6	7	8	9	10
1	Office manager	J. Smith	60	45	5	–	–	5	–	–	50	10
2	Typist/secretary	H. Jones	10	–	–	–	–	5	20	20	–	–
3	Order clerk	W. Davies	150	110	40	–	10	–	–	–	–	–
4	Order clerk	B. Robbins	150	110	–	20	10	–	100	–	–	–
9	Customer liaison	J. Molesworth	–	90	–	150	120	–	–	–	30	5
10	Order clerk	K. Barnes	150	100	–	20	10	–	100	–	–	–
11	Order clerk	R. Baines	100	90	–	150	10	30	–	–	–	–
		Totals (h)	15	12	2	5	4.5	3.2	8	8.5	4.2	1

Off-peak work factor 35%

activity and one for normal working. The peak may be monthly as in many accountancy offices, or weekly as when weekly summaries are provided for management, or daily as may be the case in order acceptance offices either when the post arrives or from mid-morning to mid-afternoon when telephone orders tend to peak.

The time period covered by the normal work-load labour balance is fixed by the periodicity of the peaks; it will depict a weekly or a monthly or a daily workload according to the processing cycle of the office.

Each cell in the labour balance represents the time spent by the person concerned in processing the relevant message. Each person has a fixed time, typically 37.5 hours per week, available for work and the labour balance summates the total time which is logged in the individual cells and expresses it as a percentage of the total time available.

Most office workers are shocked and defensive when they find how little time they spend on definable work; it is rarely higher than 50 per cent of attendance time and sometimes falls to 20 per cent. This rather sensitive point should be explained before the exercise begins or the team will be viewed with suspicion as hatchetmen. Faces can be saved by stipulating that only activities accounting for more than say two man-hours in total per processing period should be included.

An effect of expressing activities in percentage figures is that everyone will log every activity, no matter how minor, to make themselves appear to be as busy as possible. A second benefit is that comparing ratios for each person makes it possible to see where slack time in off-peak periods is most prevalent. A third advantage is that great thought will be given by the users to making each figure as inflated as possible while still remaining credible.

In turn this means that although individual figures will be inflated the ranking of individual activities will be as accurate as can be defined without a full organization and methods (O&M) study. A final advantage is that the balance shows the impact of mechanization on staff levels. It is a simple method of including all the overhead activities of office work along with the work content of direct activities.

It is a useful document for other purposes. In one instance a local manager was opposed to the proposed system (centralizing export documentation from six marketing offices) because he thought that it would be more costly. He based his arguments on an O&M study which showed the work content of individual activities. He became convinced of the proposed benefits when a labour balance showed that 'bookable activities' accounted for less than half the available man-hours; his decentralized export documentation actually cost twice as much as the O&M figures indicated.

If the reasons for the application include staff savings or dealing with increased work with existing staff, the labour balance shows which activities need to be mechanized. It also indicates where the processing bottle necks occur.

The labour balance should be completed by the end-users with minimum supervision from team members. This may cause adverse comment from analysts who think that interviewing is the best way to seek the truth. That is debatable in any case, but the aim here is not to find the truth. The labour balance is always full of misstatements, and if the aim were to streamline the existing system the labour balance would be useless. However, the aim is to replace the existing system and all that matters is obtaining the relative resources spent on each task. The DIY labour balance does this perfectly.

The team members have no axe to grind in finding certain activities labour intensive. It is the users who define their work and not the team members and this is important from the point of view of the reliability of the ranking, and from the viewpoint of involving the user in the final design. The labour balance usually takes from two to ten days to complete and it should be cross-referenced to the message list so that every document and every type of telephone message appears on both lists.

The importance of telephone entries may be significant in the final application design. For instance in the case of an order office it may be that so much time is spent in providing information to clients about the status of their orders that the proposed system should include a 'delivery order–status' set of reports which conceivably might be posted automatically to the more important clients.

The message flow chart

The first step is to identify the major messages as depicted by the column totals on the labour balance. Generally, there will be no more than five or six, which the labour balance and the consensus of user opinion will agree as accounting for the majority of the work.

A detailed message flow diagram similar to the DFD is now drawn up for each major message as it moves from desk to desk – or from section to section if it appears that a rough sectional analysis will provide sufficient detail to locate the man–machine interface.

Each message will interact with other messages so that the skeleton of the processing procedures is drawn automatically on these initial five or six message flow charts. Figure 4.1 gives an example of a message flow chart.

Once the initial charts are drawn, their activities can be checked against the cells on the labour balance. It is a matter of judgement when to stop drawing additional charts because it is not worth charting every single entry in the labour balance cells.

As a rough guide about 70 per cent of the labour balance man-hours should be depicted on the flow charts. It is not worth charting minor activities. They have to be considered later in screen design, but not in fixing the actual

Document 1. Customer order (written)

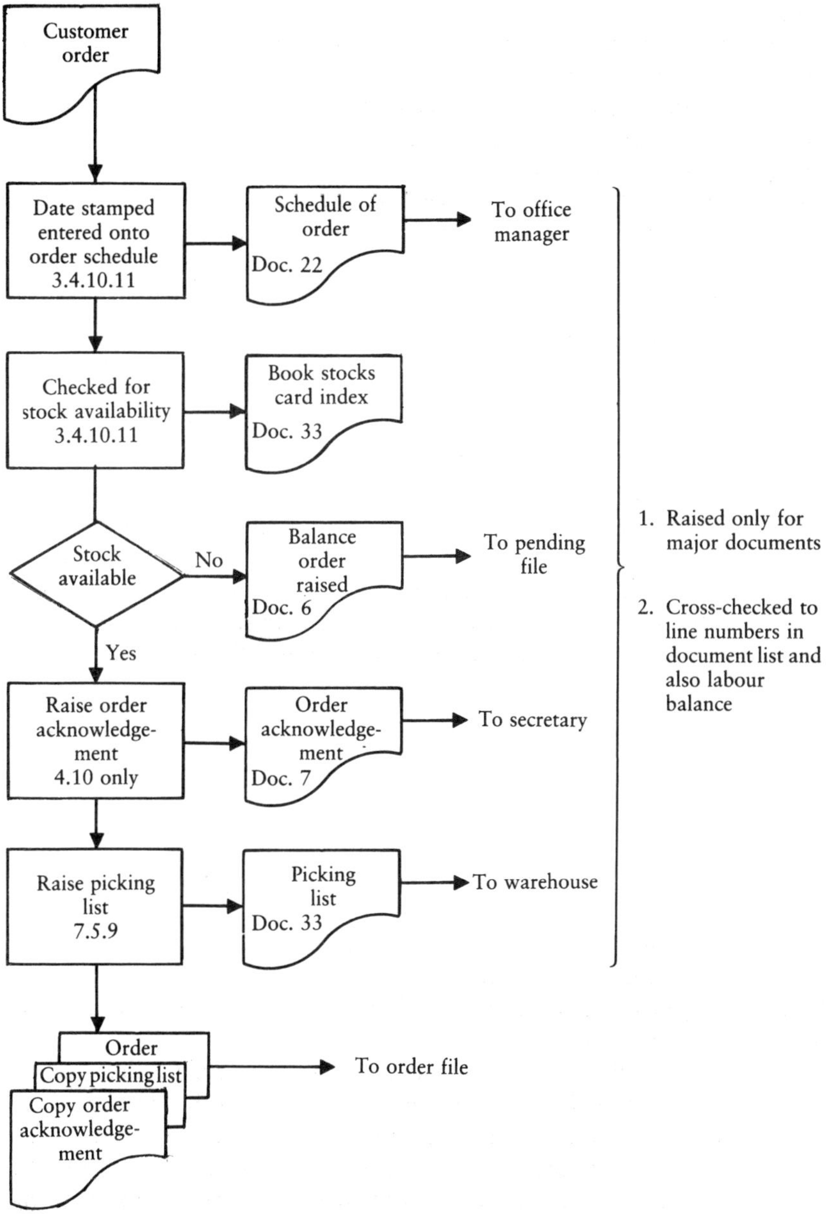

Fig. 4.1 The message flow chart.

location of the man–machine interface.

The deliverable documents so far considered are:

(a) Wish-lists.
(b) Object listings.
(c) Activity chains and DFDs.
(d) Labour balance.
(e) Message lists.
(f) Message flow-charts.

Chapter 5

Starting to Design the Application

By this stage the project team has carried out the following preliminary tasks:

1. It has determined the objectives of the application and established the reason why the application is being developed.
2. It has established a first opinion as to the extent of the application.
3. It has established the basis for management information requirements.
4. It has analysed the present system where appropriate to form a base for designing the new application.

DESIGN OF THE TOTAL APPLICATION

As mentioned in the first chapter, all these activities belong to the major development stage of defining the system in business terms and may overlap each other. It is dangerous to overlap activities from different stages but the parallel performance of tasks and activities within the four major stages should be actively encouraged. For instance it would be silly to begin the design of the computer database (DB) while still defining the application in business terms and it would be even sillier to begin thinking systematically of the application in terms of individual computer programs.

It is difficult for the experienced analyst to avoid forming opinions from the very start about the best way to tackle the application on the computer, but nothing should be written down. If there are any ideas floating around among the computer professionals they should at this stage be strictly restricted to the category of idle fancies.

If the design of the computer system is considered too early, it affects the analysis of the problem and the design of the overall system. Once the computer part of the system has been outlined the natural tendency of the analyst is to work outwards from the computer system to the manual activities rather than developing a definition of the total system.

There is a ragbag of different reasons for this. The analyst works in a department which has to master a complicated technology to produce results of any kind, so it is natural for its members to concentrate on the technical aspects of problems. Additionally at this stage the computer outline of the application seems to be the only thing that has been fixed and the analyst prefers to develop the area that has been fixed rather than engage in the examination of further hypotheses. Finally the tradition of systems development is to concentrate on the technicalities of design, because bitter experience shows that if these are mismanaged the application simply will not function.

In the majority of cases where users or management criticize the efficacy and relevance of mechanized applications in DP, it will be found that the business analysis was considered to be complete as soon as it became possible to draw the outlines of the computer kernel of the application.

The systematic consideration of the computer system design should not start until the business study is completed.

The next activity in the business study is to define the overall application more closely in terms of its inputs and outputs as this will lead automatically to a more detailed consideration of the data requirements of the application and the information requirements of its users. As usual, it is best to start by talking first to user management; to begin by talking to the users presents the possibility that management will overturn their requirements.

THE APPLICATION BOX DIAGRAM

With management, any questions and briefings should be kept as simple as possible, and the simplest way of consolidating the work done so far and taking managers to the next step is a simple box diagram (see Fig. 5.1). The manager thinks of the application as a kind of black box, so the simplest way to communicate is to draw a black box.

The box represents the total application, with lines coming into the top to depict the information which is input to the system and lines coming out of the bottom to depict the messages and information required from the system. The box diagram covers the total system, both the computer kernel and the associated manual operations which have not yet been differentiated.

The basis for defining the information, both in and out, is contained on the object listings and the activity chain which have already been initiated with management. The approximate information requirements of management for various objects have been listed on the object listings. The objects are either primary real-world objects or groupings of primary objects. Where information pertains to more than one type of object, object class or grouping (e.g. orders from customers in class 'A' for objects in class 'X'), it was listed on a 'likely requirements list' for *ad hoc* information and can be incorporated

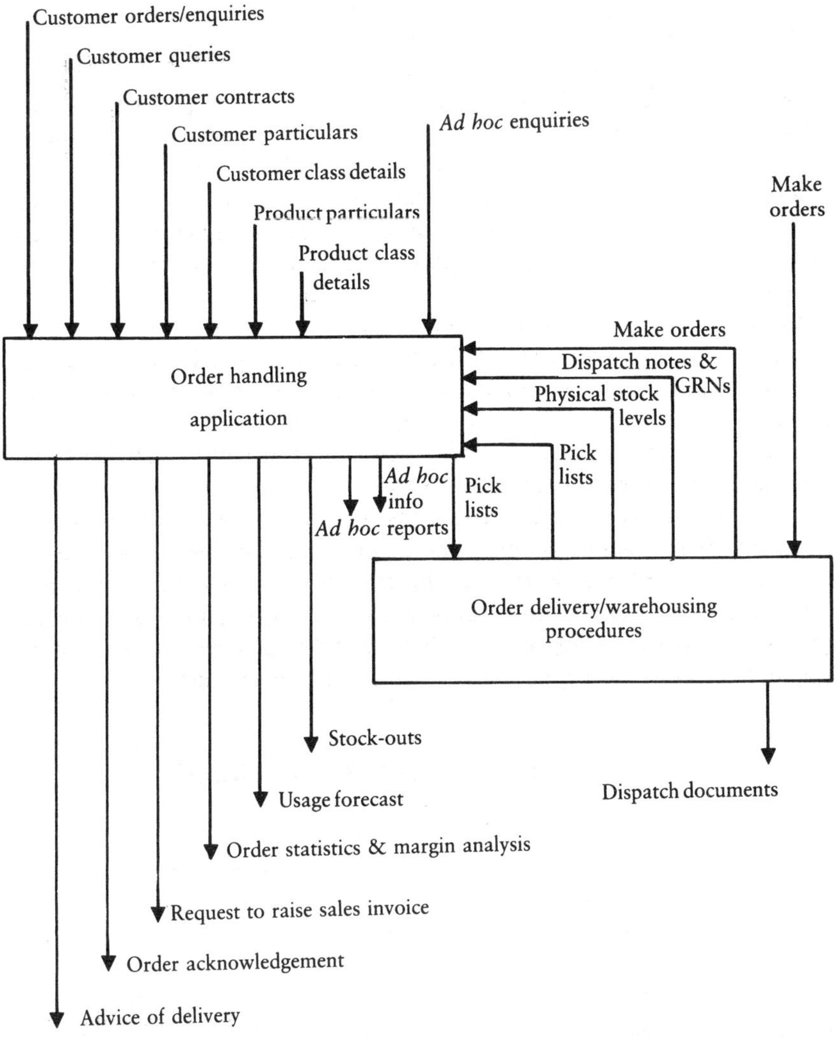

Fig. 5.1 Application box diagram.

into a single message type on the application box diagram.

A scan of the activity chain with management will automatically prompt the major types of messages needed to make the activities happen, e.g. orders, enquiries. When it becomes available, the message list produced by the end-users should be used to prompt management further on the major types of messages – usually the major messages in the labour balances.

The box diagram does not depict the actual messages into and out of the system but the total information content which the totality of messages should carry. In practice the two are nearly always the same.

The distinction between logical and physical data flows is slightly different to that demonstrated in some development methodologies. Their logical data flow is the physical data flow of a detailed DFD suitably refined to remove duplication of processing and filing, activities concerned with manual document management and matching data stores to the entities defined for the computer application. It is then defined in generic terms but still contains the same data elements as the original physical DFDs. In computer jargon it is a 'bottom-up' approach to designing the new messages where the approach advocated in this book is 'top-down'.

For instance when accepting a new customer, the two options for handling information about the new customer (name, address, terms of trade, etc.) are either that a separate message should be used to set up a customer master record, or that the customer master record should be constructed automatically from the first order which is placed by the customer. It is too early to decide which option to choose but the box diagram must contain a line that is entitled 'customer particulars', even if eventually there will be no message exclusively concerned with passing this information to the system.

The box diagram is checked for completeness against the object listings, and the object listings are updated to contain the title of the line on the application box diagram which contains the object attributes. A check can easily be made that everything on the object listings has been copied onto the box diagram. The entire procedure is designed to obtain broad-brush requirements from management using the quickest and least painful method possible. This is done by relating back solely to his original view of the application.

COMPUTER BOX DIAGRAM(S)

The application box diagram serves as the input to the activities which follow in the business study stage, which this time concern the end-users. The very first of them is to examine in detail the application box diagram and to expand and correct it wherever necessary. The end-users are the experts in the application area and it is likely that they will find marginal omissions and errors in the box diagram. There will also be omissions about activities because the manager will have considered only the major types of messages.

Depending on the magnitude of the corrections and the availability of management it might be necessary to obtain approval for the modifications. Generally it is not necessary and a note informing everyone of the changes is all that is required.

On the basis of the labour balance it is now possible to draw the boundaries between manual and computer operations in the total system based exclusively on user needs.

Business and operational requirements should not be compromised for computer convenience. The slack should always be taken up in the computer programs: the computer system must always be designed to 'get onto the users' wavelength' and not assume the contrary.

BASIS FOR MECHANIZING MARGINAL ACTIVITIES

The two factors which decide the inclusion of activities in the computer kernel are as follows:

1. Whether it is economic.
2. Whether it is technically feasible.

The economics of including or excluding particular activities is usually immediately apparent, but in marginal cases it is worth remembering that most computer applications should be required to provide a payback period of no more than two to three years. If an application costs £100 000 to develop it should give a minimum pay-off which can be safely estimated at £33 000 to £50 000 per annum.

The same rule applies to all the individual activities in the application. A manual activity costing £1000 per annum more to operate than the proposed computer equivalent should be mechanized if the programming cost of the additional programs is less than £2000–£3000. In calculating the computer operating cost at the margin (it should be at the margin and not on a full-cost basis) the additional costs are usually so trivial that they may be ignored, but the annual maintenance cost of the new programs should be estimated at about 20 per cent of the cost of writing them.

Even if the justification for the application is not direct cost-displacement, this formula for determining the economics of peripheral activities should be applied, as the more efficient the system is in cost terms the more effective it is in general business terms. A cost-effective application invariably carries with it a lot of intangible benefits.

The result of applying this type of formula is that mechanizing most marginal and peripheral activities will be shown to be economically desirable. However, they may not always be technically feasible.

This is the opposite to the views of many technically oriented experts who advise that most things are technically feasible in DP, but may be too trivial to be worth mechanizing. They are partially right in that if a problem can be defined it can be programmed, but in practice it does not work out like that.

For instance if the provision of an answer to a regular on-line query involves the manipulation of thousands or tens of thousands of records, the response time to the query may become multiples of seconds, or even minutes.

In this case it is preferable, and perhaps necessary, to design a better system either by updating a manual card index via the computer (a well-designed

Table 5.1 Example of chained log for interrogation on customer deliveries

Daily log of deliveries				27/11/85
Order	Cust.	Prod.	Prev. del.	Prev. del. cust/prod.
1234	ABCD	XYZ	15/08/85	15/08/85
		LMN		11/12/84

card index gives a response time of a couple of seconds in most cases), or by producing a list of current values which are chained (i.e. cross-referenced) to individual values in the same list or in previous lists.

Table 5.1 shows a list which 'chains' individual deliveries to previous individual deliveries to the same customer. A query about the deliveries to customer ABCD is best answered from this list without the need to spend time shuffling tens of thousands of records to produce the answer. In the example carried on the labour balance there are some two thousand delivery items raised each week, so this technique might be appropriate.

Similar situations can arise on the input of data. If indirect labour jobs are allocated to a variety of cost codes there often is no time to input all the details if a payroll deadline is to be met. In practical terms it is not technically feasible without document readers and character recognition. A labour balance would show that it is however economically desirable. One answer would be to develop a two-stage input where users were later prompted by the cost system to enter further details from the job tickets which had already been entered in abbreviated form for wages payment.

The process of deciding to mechanize or not to mechanize activities is followed for every message on the message list. It is a key activity which is not present in most development methodologies, so that when decisions have to be taken about mechanizing marginal activities they are taken much later in the development process when the prime influence is the design of the computer system. Quite often these marginal activities are tackled after the application has been developed and so contribute to the disproportionate amount of the DP budget (usually over 50 per cent) spent on so-called maintenance.

It is another example of the unfortunate impact on design of computer-oriented rather than application-oriented definitions or, in different words, or technology-driven rather than user-pulled design.

Once the application box diagram has been agreed with the end-users and a decision made on which of the activities in the labour balance are to be included in the computer system the next step is to expand the original conception of the computer system into computer subsystems. In passing, there is not the time to discuss everything with all end-users. This is discussed in the following chapter.

NOTES ON BATCH AND TRANSACTION PROCESSING

All significant DP and MIS applications have a transaction processing component and a batch processing component. In other words some work is processed as it occurs, some is done daily, some weekly and so on. The messages for each subsystem are taken from the labour balance, where volumes and timings are indicated.

Thus the application box diagram is first split into a computer and a non-computer component and then the computer component is split into on-line and periodic or batch processing. Again the split into batch and transaction processing is made on the base of user requirements and not computer convenience. This is shown in Fig. 5.2.

In some cases it may be worth first setting up a global box diagram for the entire computer system and then splitting it into subsystems. This is generally a waste of time. Most analysts are able to deduce which messages should go into which subsystem from the application box diagram and more detailed explanations from the users.

The default choice should be that everything should go into the transaction processing box unless the work flow in the office or the amount of required data manipulation demands that the activity be carried out in batch. Batch programs are more costly to write than transaction programs, and batch processing in an office is usually expensive, as the batches need to be controlled. In addition there are administrative overheads in filing.

A compromise is to process transactions as they occur and to store the output on disc until required, when a batch of documentation may be produced. One example is the production of advice-of-delivery notes on special stationery. To avoid changing stationery each time that advice notes are produced, they can be kept as print images and run off in batches at intervals. This type of transaction is treated as a transaction process because the programming is exactly the same as for a normal transaction.

THE TURNAROUND DOCUMENT

The 'turnaround document' is a powerful and neglected tool for DP and sometimes for MIS applications. Its range is wider than might be imagined.

On looking at the computer box diagrams it is seen that in some cases a message is produced by the computer system either for further manual processing or to trigger some sort of manual activity (e.g. the pick lists). On completion of the processing or activity another message is then passed back to the computer system informing the computer of the results. This gives the opportunity to make provision for turnaround documents. The use of activity chains in the initial requirements analysis leads naturally to the consideration of their use.

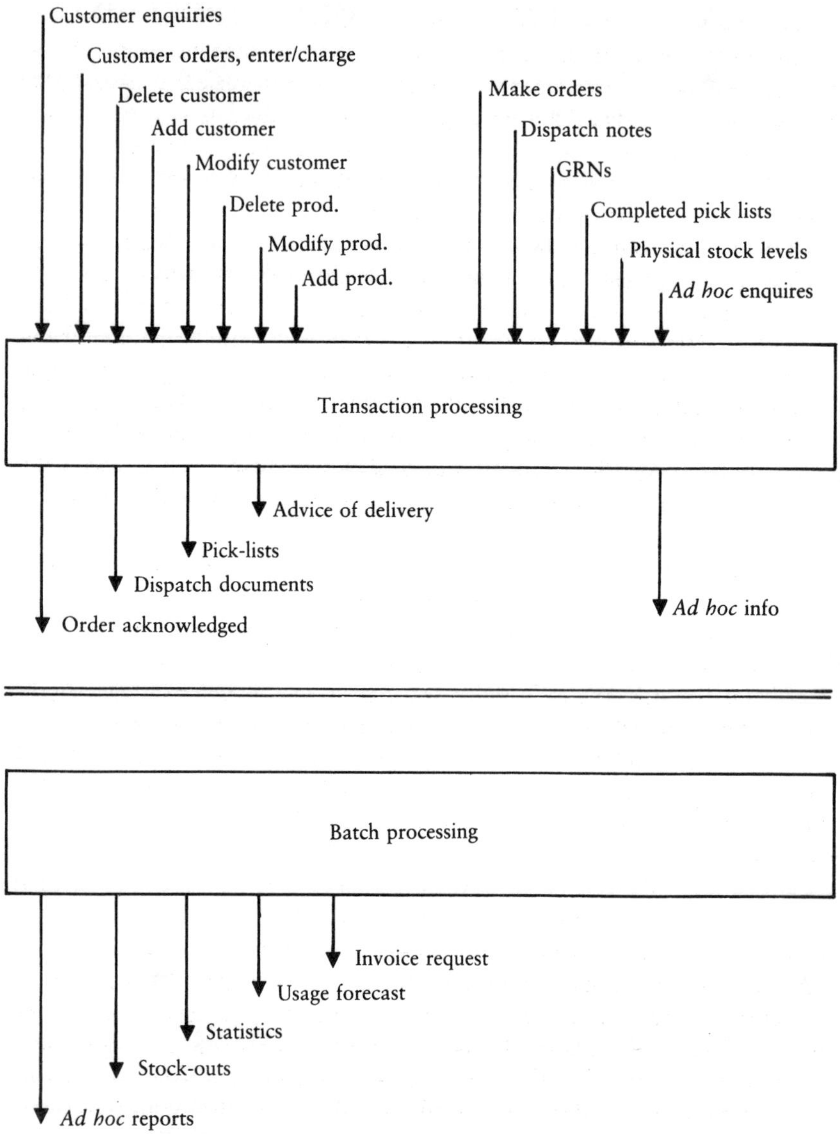

Fig. 5.2 Transaction and batch processing.

A common example is the allocation of cash receipts to outstanding invoices. The usual procedure is for an accounts clerk to allocate the total amount of the payment received to a list of outstanding unpaid invoices. This information is then passed to the computer system so that the computer DB may be updated.

One way of handling this procedure is for the computer to provide each month a full listing of all the accounts and their outstanding payments. When the cheque arrives the clerk allocates the monies by merely ticking lines on the computer listing, and perhaps writing a new line for a balance amount when the amount received does not correspond with the outstanding debt.

The input to the computer need now only indicate the account number and line numbers, including that for the new balance amount. The computer can now balance the payment with the amounts on the lines indicated, and can produce a new listing to update the manual file of the accounts clerk.

In this case the work of data entry to the computer system has been reduced, and the possibility of clerical error has been eliminated by a computer check on the arithmetic of the accounts clerk. It is a good example of how cost-effectiveness brings other benefits. It also leads by a natural progression to using the computer to actually allocate cash and provide merely an exception list to the accounts clerk for those receipts which the computer cannot allocate. This is a substantial saving when scores of payments are received each day.

An example in the order-book application is that the acceptance of a customer order triggers the production of a pick list for the warehouse. The computer should keep a copy of the pick list so that when the list returns to the computer terminal from the warehouse to show what products have been picked for dispatch, the only input required is the document number previously allocated to the pick list by the computer. If a line number is also typed on input of the completed picking list, this indicates that the line was excluded from the order, and if a line number followed by a quantity is typed this indicates that only the quantity typed was dispatched and a balance order is necessary.

The turnaround document can be used wherever the computer can deduce what should result from the information which it places on an output document (e.g. an invoice paid, a product is picked).

A public utility produces thousands of invoices every day. They are based on meter readings submitted to the computer to trigger the production of invoices. The meter reading sheets, if they are not already computer produced, can be produced by the computer as a by-product of the invoicing run for a cost which is no more than the additional cost of continuous stationery and a few units of electricity. (Printing takes up practically no time on the central computer so there is rarely an opportunity cost to producing computer printouts.)

At this stage the alert analyst asks if there is the possibility of a turnaround document, and there is. The computer can estimate the likely meter reading for each address on the list and can print for each meter a minimum and a maximum reading. If the actual reading falls between the two then the meter reader need not do anything except tick the document to indicate that he has read the meter.

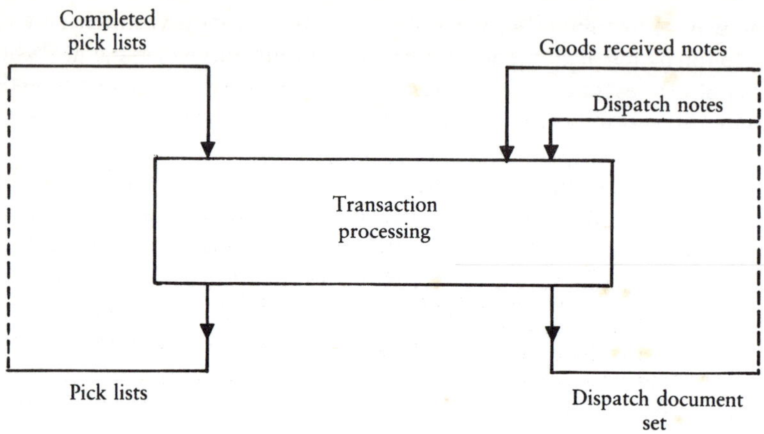

Fig. 5.3 Box diagram with turnaround documents.

Exception reporting can now be combined with the turnaround document. Only those documents where the reading was outside the range calculated in advance by the computer will be input to the computer. This removes a major operational cost for the billing system.

The reader might ask why time is being wasted on such obvious examples of computer utilization. The reason for emphasizing such banal examples of user-oriented design techniques is to emphasize that the DP tradition misses many design opportunities by skimping the business analysis. The author has lived in North America and several European countries and has never seen the above technique used by any public utilities.

The best way to ascertain possibilities for turnaround documents is to take the initial activity-chain list (or DFD if there is no list) and to produce a more detailed list of activities in conjunction with the end-user. Once the activities in the chain have been identified the possibility of each activity being triggered by documentation from a previous activity should be systematically considered.

Where possible, the computer box diagrams can be modified to show likely turnaround documents (see Fig. 5.3). Every document should be considered as a possible means of increasing the efficiency and reliability of the system.

LISTING THE MESSAGE CONTENTS

Until now we have listed the information content input and output for the total system and for the individual components of the computer system. The messages themselves have not been listed.

In most cases the messages correspond exactly with the grouping of information but occasionally they do not. The final activity on the definition of the system at the operational level is the listing of the contents of messages.

First it is assumed that all messages correspond to the grouping of information, e.g. if there is an order there will be a message for it, likewise for an enquiry and for a cancellation, etc. The likely messages are listed and categorized under the following headings:

1. Operational administration (orders, pick lists, etc.).
2. Exception messages to operational management (stock-outs, overdues).
3. Information messages to management (trends, summaries, comparisons).
4. Requests for interrogation from everybody.

The categories indicate the possible interactions of messages which are generally limited to messages within a category. A stock-out message might be triggered by the processing of a pick list, but it is not an inevitable result as is say the production of advice-of-delivery notes. The categories ensure that possibilities (e.g. turnaround documents) and user needs for all kinds of messages are considered systematically rather than on an *ad hoc* basis.

Each category should be considered in turn and every possible message should be included during a series of free-ranging discussion sessions. The resulting message contents list should be compared to the computer box diagrams. The discussion is a further check that nothing has been omitted in the previous systems definitions.

Once the message contents list is complete the messages should be 'commonized'. This is shorthand for saying the different messages should be merged if possible to make the resulting system more simple and more robust. For instance an order enquiry and an actual order might contain substantially the same information, and the analyst can then decide that he should use a common screen and a common program for the two messages, with an indicator field showing which is being processed.

There will be fewer programs to go wrong, fewer kinds of documents to file and so on. The fewer the varieties of message the less will be the operational overheads of the system both on and off the computer.

The procedure is to take every information group from one message type in turn to see if it can be included into another message type. The possibility of merging the messages for setting up a new customer and setting up a customer order has already been mentioned. Other possibilities are merging the messages for deleting orders and amending orders and raising balance orders, and so on. The aim is to minimize the forms required and the variety of different procedures in the eventual system.

Once this has been done the results are mapped onto a final computer box diagram (see Fig. 5.4). The document is very easily and quickly produced and it now gives the final statement of the application in operational terms.

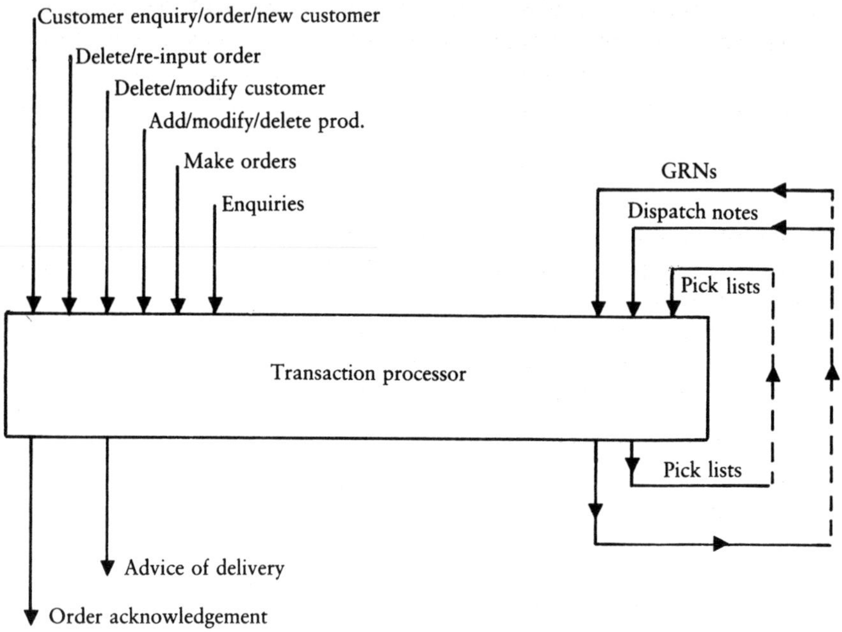

Fig. 5.4 Simplified final computer box diagram.

THE FEASIBILITY STUDY AGAIN

The business definition should take no more than eight or ten analyst weeks. This may or may not be sufficient to produce a full business definition with the detail to form a base for the design of the new system. The DP analyst usually will be able to estimate before he starts whether this is likely.

If eight or ten analyst weeks is not sufficient to produce the full business study, there should be a feasibility study in order to limit the cost of finding possibly that there is no case for proceeding further, and to improve management control by reducing the time reporting span of the team.

The feasibility study would define the system in its essentials and not attempt to define the application sufficiently to serve as a base for technical design. The object listings may not be entirely complete, or there may be loose ends on the box diagram, etc. The labour balance and message lists might only account for say 50 per cent or so of the time available.

However, the feasibility study should not compromise on dealing with the wish-lists, activity chains and associated DFDs. It should take one to eight analyst man-weeks. If it takes less than one man-week a feasibility study is probably not required.

If a feasibility study takes more than eight or ten man-weeks the project is too large for the normal rules to apply. This might justifiably be the case say for major banking applications or major government applications. However, the acceptance of a project as too large for an eight or ten week feasibility study should be initially rejected by management. The team should be asked to split the application into a series of smaller applications which can be defined in the normal timespan.

The attitude by management should be that this is always possible until the reasons for it not being possible are listed and defended. The reason is quite simple: they are being asked to accept a capital expenditure of at least several hundred thousand pounds and probably over a million pounds, which will be hard to manage. They should ask exactly the same questions that are asked when similar expenditures are mooted in other areas.

At this stage the application has been defined using:

(a) Application box diagrams.
(b) Computer box diagrams.
(c) Message content lists.
(d) Turnaround documents.
(e) Perhaps a grid for listing information elements on individual messages (i.e. the message contents list).

These documents were completed by considering information on previous documents concerned with the business and operational aspects of the application, e.g.:

(a) Wish-lists.
(b) Activity chains (DFDs).
(c) Object listings.
(d) Message lists.
(e) Labour balance.
(f) Message flow charts.

In particular the border between the computer kernel of the application and the total application area has been defined and justified in such a way that the business objectives laid down by management and the operational objectives of the end-user have been fully taken into account.

Chapter 6

User Participation

JOINT PROJECT TEAMS

It is apparent from the previous chapters that the development of a DP or an MIS application is a process that should marry the technical skills of the computer professionals to the application knowledge of the end-users and to the knowledge of the overall organization and its environment which is possessed by management.

It follows automatically that the team to define and construct the application should not be made up exclusively of DP staff. It is a pity that this fundamental truth is not emphasized more strongly by DP management and by management consultants. There are many organizations where only lip service is paid to the idea of joint development. The development team consists exclusively of computer technicians, often only recently hired by the company, who try to explain to the users and to management what they are doing as the exercise progresses. It is like a whale trying to talk to an elephant.

This lack of user involvement is puzzling as it was identified as a problem almost from the start of business computing. In 1961 for instance, McKinsey management consultants published the results of a survey. 'Unlocking the computer's profit potential', which clearly demonstrated the correlation between effective systems and user involvement. In 1971 a survey made by the British Institute of Management advised 'stop looking at computers and start looking at your business, do not regard the computer as a panacea for the problems of your organization'. In 1985 a book advising managers about computer applications (*Ending the computer conspiracy* by Colin Corder) commented 'First define the problem . . . The end-product is seldom to the user's complete satisfaction . . . much of the art of problem solving consists in arriving at a clear definition of what the problem actually is.'

It can only be hypothesized that the traditional career structure in DP of moving into analysis from programming means that each generation of analysts has to reinvent the wheel.

The development team should always contain members from the application area and should always be able to obtain an input from top management in a systematic fashion.

The more time that the end-user can spend on helping in the definition and the construction of the application, the better will be the final product. It cannot be said too often that end-users are the only people with the necessary knowledge of operational detail, and management is the only group with the necessary knowledge of the business. The major reason that most applications fail to deliver benefits which are there for the picking is the lack of user involvement in the development process.

WHY USERS ARE NOT INVOLVED

The most common reason for end-user exclusion is the lack of understanding of the need for end-user involvement on all sides; the computer department, the management and the end-users themselves do not seem to realize that the knowledge about the application area which sometimes has taken years to acquire cannot be passed over to an outsider in a few days or weeks. In most cases the end-users are not represented on the development team because the computer department has not insisted on their participation.

There are however instances where the computer professional tries strenuously to obtain end-user involvement and fails. In some cases the reason for mechanizing or further mechanizing an application area may be that the present procedures and all the staff concerned with them are overworked. It may not be possible to deplete further the operational staff by seconding some of them to the development team.

Another reason may be that the application is intended to provide a substitute for experienced staff. It often happens that two or three key personnel retire, or give notice all at once, and the application area finds itself without a core of expertise.

Yet another reason for the lack of end-user expertise in the development team may be that the activities in the application area have changed substantially in a short time. Perhaps the company has been merged with another organization, or a new market has emerged, so that the traditional means of administering the traditional market no longer suffice. Whatever the reasons, the application expertise which normally exists is absent.

However, in most cases a lack of application expertise in the development team is simply because it is not perceived as necessary. If the team takes every opportunity to find expertise it is usually available. My method when I was a DP manager was to refuse any request for development unless user expertise was made available. This always seemed to work.

GREEN FIELD APPLICATIONS

A variant which is more common with MIS projects is that the application area is 'green field'; the application does not replace something that already exists, but aims to provide facilities that are entirely new by building the application on a green field site.

With green field projects, where by definition there is no user expertise, the danger lies in not thinking through the implications of the user or management requirements. Invariably this means quantifying what is involved and asking questions like 'how much information?' and 'when will it arrive?'. The analyst has to operate on Faraday's tenet that nothing can be understood unless it can first be counted.

In the early days of computers a common development problem was that managers reacted wrongly to the computer's ability to provide information cheaply and quickly. They were not yet accustomed to the idea of *ad hoc* information on demand and asked for everything on a regular basis.

In one case a production manager insisted on daily information presented in dozens of different permutations regarding the detailed performance of twenty machines on processing some hundreds of individual operations. He was only dissuaded when the analyst pointed out that a five hundredweight van would be needed to transport the daily listings from the computer centre to his office.

In another case the sales managers of eight regions demanded detailed sales statistics on a national basis and combined in every conceivable way. Some 30 000 individual sales items per month would have produced two million lines of analysis all of which would have been duplicated eight times.

In green field applications the only partial answer is to insist that everybody learns to quantify their requirements in order to provide a sporting chance that the new environment can in turn be quantified and defined. This is not an easy lesson to teach to managers. The development team has to guide management on the amount of information that it can realistically absorb on a regular basis. As many requirements as possible should be defined as *ad hoc* requests, they can always be regularized later. The converse is not equally true.

If the advice sounds banal it should be pointed out that most major changes in public administration here and abroad have had inordinate teething troubles because it was ignored during the initial stages of development.

Ideally the development team should have an experienced and bright person from the application area seconded to it. This implies a fairly senior person, say a section head or senior professional. When additional knowledge or end-user resources are required it should be the responsibility of this person to obtain them. Unfortunately, if the person is experienced enough to be part of the development team he or she is usually too valuable to be spared.

A compromise is a part-time secondment, and if that is not possible a senior person in the application area should be nominated as the contact for the team. This means that any query from the team should be tackled within an hour or so of the query arriving. If even this is not possible it is wise to ponder the desirability of proceeding further.

Some activities of the seconded user have already been seen, and others will emerge as the development of the application unfolds.

REASONS FOR USER INVOLVEMENT

There are two reasons for insisting on this formal user participation. Firstly, the team gains the application expertise which will be necessary if anything better than a barely adequate application is to be developed. Secondly, the end-user contact becomes a means of dealing with suspicion and unease in the end-user department.

The end-users know that the team cannot hope to grasp the finer points of their work and suspect that the eventual system will suffer because of it. If one of them is seconded to the team their suspicion largely disappears. This is especially the case if the user is obstinate in his perception of the advantages of the present system. Once he realizes that these perceived advantages can be designed into the new system he will become an enthusiastic advocate of change. The other users now have a point of contact with the team. The seconded person can deal with queries on the spot and ensure that the development team takes them into account.

It is a good idea to allocate desks to the team in the application area whether it is a green field or a replacement application, even if the work could be done just as well in the DP/MIS department. What must be avoided is the impression of outside boffins not considering the effect of their designs on the people who will have to operate them.

MANAGEMENT INVOLVEMENT AND CONTROL

The team should only expect limited involvement by management and should plan accordingly. Just as the involvement of actual users is not stressed enough by DP gurus, the desirable degree of management involvement is overstated. One wonders sometimes in DP textbooks if the MD is there to run the company or to act as longstop to the development team.

The probable reason is that most DP gurus are consultants who have access to higher levels of management than do in-house staff for any application. The very fact that an outside consultant is involved means that the application has a high profile and management will take a deeper interest in it. More mundane applications, the majority, have to be satisfied with less management time.

The job of management in systems development is to provide information which only they can provide, and to approve the major design decisions in view of their general business implications.

Time spent with senior managers has to be carefully planned in advance and the interviews themselves must be prepared to a format. The DP jargon is that the interviews must be structured. In the examples given in this book and with most development methodologies, the interviews are automatically structured by the documentation which managers help to complete.

PROJECT STEERING COMMITTEES

A project steering committee should be set up for every project to ensure that it runs smoothly along the lines laid down in the initial stages. A senior manager with responsibility for the application area should chair the meetings of the project steering committee. For instance in a general ledger application, the chief accountant, or his principal lieutenant, should chair the committee.

The committee should meet at most weekly, and at least monthly, for the duration of the project. The time for meetings should be fixed in advance. The committee is an advisory committee to the executive chairman. He would be foolish to override the advice of the committee, but once or twice such a situation will arise. In most cases who takes the decision is not an issue; the required actions are invariably so plain that decisions take themselves.

It is permissible and desirable for the computer members of the committee to lead the chairman. One way is to make the development team leader the secretary, as then he controls the minutes. These can be written to provide the emphasis which he thinks appropriate without diminishing the chairman's authority. Figure 6.1 gives an example of the minutes of a meeting. This eliminates ancillary briefing papers and serves the same purpose. The ground rules should be explained to everyone at the first meeting.

A minute should always indicate an action and a person responsible for carrying it out. In this way much development documentation is replaced by

MINUTE NO.	ACTION
78. The telephone log of back-up messages to the warehouse needs more detail. The split between progress queries and stock queries is not present. A sampling operation should be set up to run over a five-day period.	RWT
79. Information requirements for the classes of customers is incomplete. Meetings should be set up with marketing, company planning, and cost-accounting.	JCD

Fig. 6.1 Example of minutes.

committee minutes. The chairman automatically becomes more involved in the project as he is probably quite familiar with exercising control in this way.

The core of the steering committee should be not more than four or five people. There is usually the chairman, the secretary, the manager responsible for the operational running of the application area, and the senior technical person from the development team. From time to time additional people may be co-opted as their portion of the project, on the user side or the development side, comes under the microscope. But there should never be more than six or seven people attending a meeting.

Committee meetings should be scheduled for an hour or a maximum of two hours. These are steering committees and not operational control bodies. The operational control effectively reports to the committee and gains approval for the next stage.

The system works well as it brings together both sides on a regular basis. If for instance the team want to request a labour balance when the application area is processing a periodic peak load, this can be explained in advance. If tasks are running late for whatever reason, the problem can be highlighted before it starts to cripple the project.

Finally, the minutes should be written in terms of the activities required to produce deliverable documentation. The deliverables discussed up to this point are:

(a) Wish-lists.
(b) Activity chains (DFDs).
(c) Object listings.
(d) Message lists.
(e) Labour balances.
(f) Message flow charts.
(g) Application box diagrams.
(h) Computer box diagrams (different levels of detail).
(i) Message content lists.
(j) The optional grid for analysing message contents.

Chapter 7

Choosing the Right Development Option

OPTIONS ONLY RECENTLY AVAILABLE

Until the early 1980s, choosing the best of the available development options for an application was an unavailable luxury, as usually there was a central corporate computer and the only choice was whether or not to increase its size for a specific application. Similarly, the application would always be developed by computer professionals within the rigid framework of standards and facilities established by the computer department.

Until the advent of the microcomputer (or PC) and the blossoming of the minicomputer, choosing an option for developing and operating an application was a case of choosing 'any colour so long as it's black'.

Most writings on systems analysis and design are still based on this traditional past. They assume incorrectly that the normal method of systems development is to develop a set of specifications for a set of handwritten programs to carry out the task defined in the business study. This is still a possible option for most applications, especially in large organizations, but it is no longer the only route to the design and implementation of a DP or MIS application.

NEW DEVELOPMENT OPTIONS

Besides the continuing advances in hardware power and speed and the vertiginous fall in hardware prices, there have also been less heralded advances in software. In particular, the arrival of *fourth-generation languages* (4GLs) and productivity tools for the analyst (which go by the name of analyst work-benches or confusingly by the name of 'computer aided software engineering' (CASE)) has affected the development of even traditional applications.

In parallel with these new tools for the developer there has arrived a new generation of application packages which are both more powerful and more flexible than their predecessors. An increasing proportion of these packages incorporate the advances made in language design so that they can be easily tailored to meet a much larger variety of individual requirements.

Finally a completely new kind of software has been developed which is designed to be used by end-users directly. These are packages and languages and application packages which are expressly designed to be utilized by computer nonprofessionals.

In larger installations it is normal to find that a special centre has been established to guide and encourage the use of such tools directly by end-users. Sometimes called an 'information centre' sometimes a 'user support centre', the function is always the same – to encourage the use of 'do it yourself' development tools by the end-user with advice and guidance from the computer department.

The options now open to the development team include the following:

1. Traditional development of custom-built programs.
2. Procurement of an application package which can be modified if necessary by the development team.
3. Partial or full end-user development of custom-built programs using fourth-generation technology (4GT).
4. Partial or full end-user development of a system using end-user tools.
5. Utilization of end-user productivity tools for use in an existing system (e.g. spreadsheets, electronic mail, graphics, etc.) in lieu of developing a completely new system.
6. Any combination of the above.

The first activity in the development of the computer kernel of the application is to decide which of these options to pursue.

The only people qualified to evaluate all these options are the technical members of the development team, but they are under no pressure to do so, and often have vested interests to protect. Not surprisingly the decision is often taken by default to develop the new application as a traditional type computer system using traditional development methods. The only people who can insist that the evaluation of alternatives takes place are management. The implications of choosing one option rather than another have much more to do with business structure than with technology. Unfortunately managers generally do not know enough about state-of-the-art options in computing and fail to insist on a proper evaluation.

There is no rigid set of rules which can be applied to select the best option as organizations and applications are all unique, but the same factors have to be considered in each case. In different situations the same conclusions from an examination of alternatives can lead to selecting different options because the selection is dependent also on nontechnical considerations.

For instance an application to consolidate year-end accounts from subsidiary companies might easily be demonstrated to best fit the purchase of a powerful multi-user micro providing a spreadsheet with communication links to mainframe DBs. This might be unacceptable however to management because it might conceivably allow a degree of discretion on the part of the consolidator, which the chief accountant would veto. He might insist that the figures he consolidated by custom-built programs on a corporate computer.

The likelihood of different choices being made in different corporate situations from the same set of technical conclusions does not invalidate the need to carry out the analysis. The principal factors to be taken into account are:

Availability of application packages

If the application is one for which there are a multiplicity of packages available on the market, they should be examined. This applies to all financial accounting applications, including receivables and payables as well as inventory control, production planning and purchasing. The packages need examining to ensure that they are sufficient and technically sound and above all have features to communicate with other applications on the corporate computer.

MIS and DP applications

There is no absolute rule, but it is more likely that the skills of the computer professional will be needed for the development of DP applications than for MIS applications. DP applications have to be closely integrated with other applications, either mechanized or manual and to do this adequately demands experience-based skills.

Additionally DP systems work is a full-time job. The end-user is unlikely to possess either the skills or the time to develop a nontrivial DP system.

Updating the corporate data base (DB)

Another factor is that an MIS application rarely changes the application DB, where a DP application always changes the DB. Most company information originates in the raw data of DP applications.

In most cases free access to modify a corporate DB is too dangerous an option to be given to the end-user. If the DB is inadvertently corrupted the organization in many cases cannot continue to function. If the DB serves more than one application, it is also unlikely that the DP department will allow the end-user to decide how and when the DB should be updated, because of the knock-on effects for other applications.

Interrogation of the DB is a different matter. Provided that adequate security prevents end-user access to sensitive information, the DP department usually views this with approval. It poses no operational danger and it removes much programming work which DP professionals find both trivial and irritating.

There are some departmental DP applications which despite updating the DB can be considered for end-user development, if it is technically and managerially desirable. Examples might include allocating service charges in research projects or engineering maintenance, analysis of time sheets, processing of expense claims, etc.

There is additionally the possibility that end-users might be utilized to develop some of the programs in an otherwise traditional development application. They might for instance write reports and design interrogation screens. This option will become more usual when DP realizes that the core of an application often accounts for only about 40 per cent of the code. The remaining code can be farmed out to end-users operating under the functional control of DP.

CHARACTERISTICS OF END-USERS

Managers and clerks generally do not have the time to develop either MIS or DP applications. Even if they had the time it would not be possible for them to give the work the priority it would need for satisfactory completion. It is however possible for both classes of users to be provided with simple tools to write simple programs, perhaps by extending an existing system.

Ad hoc reporting and information retrieval aspects of a system can be developed for all classes of end-users on a formal traditional basis, or left undeveloped and designed as and when required by the end-user using fourth-generation technology (4GT), (e.g. report writers, enquiry languages, etc.).

Degree of integration with other applications

The more intimately an application is meshed with other applications the more likely it is that the solution will be a custom-built system designed by DP professionals. This becomes even more true if there is a communication component in the system.

The messages between applications have to be checked to see that they do not cause the receiving system to malfunction and to see that they satisfy corporate requirements for the validity of widely used information. The further reasons for using the DP professional in this type of application lie in the technical difficulties of establishing links between complex applications.

Volume of traffic

The application processing cycle is a concept used to indicate the processing between the first time that the first program in the cycle is switched on to the closing of the final program in the application for the last time. There are daily, weekly, monthly and quarterly processing cycles which are set up on the computer, operate on messages, and then are closed prior to the commencement of the next cycle.

If the application deals with hundreds or thousands of messages per processing cycle, it is almost certainly an application for development by the DP professional. If there are no complicating factors such as very complex processing or complicated data structures, a few score messages per cycle open the door to the possibility of end-user development, but the balance of probability still lies with the professional DP development. By the time the volume has dropped to one or two dozen messages per cycle the possibilities are evenly balanced. A handful of messages per processing cycle implies that, if the other factors mentioned permit, it is probably an application for end-user development.

The reason has to do partly with system efficiency. End-users generally develop very inefficient systems which demand perhaps a hundred times more machine resources than would a professionally developed system. This may be acceptable for a handful of messages per day but would overload the computer if the volumes were high. The inefficiency of end-user developed applications is due to inefficient 4GT software, and to inefficient and amateur design.

Another reason is that a high volume of messages generates problems of control and security about which the end-user knows very little.

Nature of the application

If the labour balance shows that a significant proportion of time or the critical activities in the application are concerned with calculation, the option exists for providing productivity tools rather than a completely new application. The same option exists if the labour balance shows a signficant amount of time is spent in processing text. The higher the proportion of calculation and or text manipulation in the labour balance the more likely it is that productivity tools should displace, at least partially, the traditional systems type solution.

Typical examples might be the customerization of standard manufacturing routes for particular orders, export invoicing, some costing exercises, technical quality control, etc.

Conversely, the more time that is spent in retrieving, transmitting, filing, and comparing information the more likely it is that a traditional DP system (either custom programs or packaged, either end-user developed or professionally developed) is the answer.

Obviously the more time that is spent on retrieving information and summarizing information the more likely it is that the application will be an MIS application, constructed via the same options as for DP.

CHOICE OF MACHINE

Once it has been decided that the application is probably:

(a) A packaged system;
(b) A custom-built system;
(c) End-user developed;
(d) Professionally developed;
(e) Productivity tools substituting all or part of the proposed system;
(f) A mixture of the above;

the next step is to choose the machine.

The choices are as follows:

1. To centralize or to decentralize using some kind of network.
2. To use existing equipment or purchase new equipment.
3. To design everything as transaction processing or to develop a batch element, e.g. either to produce invoices in a batch at the end of the day, or to produce an invoice at the end of each invoicing transaction.

Factors to take into account in machine selection are discussed in the following paragraphs.

The trend to distributed systems

In the last few years there has been a tendency to develop applications whose processing and filing capabilities are distributed over more than one machine. The machines may be a mixture of mainframes, minis and micros.

The initial decision should be to make the system as technically distributed as possible unless good reasons emerge for the opposite approach.

The reason is that demand for computing is expanding faster than the manufacturers can increase the capacity of mainframes. In most organizations it would today be impossible to carry out all IT work on mainframes, they simply could not handle the work carried out on dozens of departmental minis and hundreds of micro machines. The rate of increase in demand for computing is still accelerating and while the computing handled by mainframes will greatly expand in absolute terms, it will fall in relative terms.

Also the law of diminishing returns applies as machines get bigger. As machines become cheaper the economies of scale diminish, but the non-machine overhead costs of using them are unchanged. Other things being equal, a networked application will have a long life, and will be on a long-term cost curve that is dropping, and will be open to future development and extension. The opposite applies to centralized solutions.

Batch or transaction processing

These days nearly all systems are transaction based for data input and updating the DB. The question of choosing between batch and transaction processing usually only arises for other functional activities such as the compilation of derived information and the production of regular output listings and documents.

The logic of batch processing is more difficult to design than the straight transaction processing of a single message, in fact many 4GLs do not offer batch facilities because it is too difficult to provide generalized logic. Outside the computer, most information handling is transaction based, so that the design of a batch system on the computer will usually put additional work into manual procedures. Batches of work have to be assembled, scheduled and controlled. Even if operations do need to be batched in the manual system, it is always possible to batch output at the last minute from a transaction based computer system. The output from a transaction is placed on disc and then sorted and presented in batches as required. The computer system looks like a batch system to the user but it is not.

However, batch processing is much more efficient than transaction processing for large volumes of transactions, and if the volumes are very high (gas or electricity billing) it may be the only answer. This is because information can be fetched in large parcels from disc instead of the penny packets used by transaction processing. If most of the data in a DB file is processed during a processing cycle (e.g. there is a high 'hit rate' as in payroll) the operating economies of batch are maximized.

The economic advantages of batch systems are steadily being eroded by cheap machines, but the inflexibility of batch systems remains unaltered both on the machine (difficult to extend and modify) and off the machine bureaucratic document control procedures. The default decision should be to design the system as a transaction processing system, i.e. leave it transaction based until forced to change.

To purchase more equipment

If it is decided to purchase additional equipment to provide more capacity or because of its special features, it should be noted that users should not be asked to operate a variety of equipment for different applications. With a productivity tool-cum-DP type of application both the productivity tool (e.g. spreadsheet) and the system should operate on the same terminal. The user does not want one machine for word processing, another for personal computing, and yet another for corporate computing. He wants a general purpose terminal. This implies opting for a micro workstation instead of a dumb terminal, at least in the long run. In the long run the dumb terminal will be limited by the same factors which limit mainframe potential growth.

BASIC SOFTWARE

The final step in the process of choosing development options is to evaluate the functional software associated with the shortlist. For instance one might need to evaluate local area networks and the way in which they interface with the hardware and development software (packages, languages, tools) already discussed.

THE SELECTION LOOP

The selection loop now becomes:
Evaluate in turn the following:

1. Type of final system (tools, packages, customized programs).
2. Type of development (end-user, professional DP).
3. Type of application architecture (centralized, distributed).
4. Type of functional software.

At each step return and recast the previous step. At the end of the list go back to the beginning to check that the choice of options is capable of providing the right answer.

The selection loop works its way down to the technical criteria for choice within a framework of business criteria. It thus neatly turns upside down the usual approach to deciding first the technicalities when selecting options. It is probably because the DP tail wags the business dog in selecting options that many computer applications provide suboptimized benefits.

PROTOTYPING

4GT and especially 4GLs have so reduced the work of writing programs that they automatically revolutionize the whole of systems development. The traditional DP method aims to produce consistent specifications for a set of computer programs. The costs of programming were historically so high that tight technical specifications were needed. Controlling the production of these specifications became a major objective in systems development.

With 4GLs it is often quicker to write a program directly than it is to write technical specifications, and if the program does not suffice it can be scrapped and a new program written. Sometimes this 'prototype' is used as the production program, and sometimes it is used merely to demonstrate the system to the end-user.

It is also possible to prototype parts of a system, for instance the screens and the reports, while still using traditional methods for the remainder. It is difficult to find applications where the use of 4GT of some sort is not apposite.

The use of 4GT changes the detailed designing of the system. Prior to 4GT the system had to be defined in absolute detail before programming could be permitted to start. It was the only way to write consistent programs. Unfortunately this assumed that the user was able to define the system in absolute detail without ever having seen it in action.

This assumption has been known to be false for years, at least by DP managers, but there was not much that the computer professional could do about it. Unfortunately DP managers, who have to live with end-users, are rarely technical innovators. They do not have the time. Technical innovators, who are rarely exposed to end-users, took longer to realize the situation. It is interesting that the innovators who carried the banner for relational databases (RDBs) and for 4GLs and data driven design (Codd, Martin and Jackson), spent long periods on actual application work.

With modern tools the analyst is able to produce prototypes while working closely with the user. The user can now actually use a screen before accepting it. He or she can request *ad hoc* reports from a prototype system and can find the weaknesses in their original design. The design emphasis is shifting from being driven by technology to be pulled by user requirements as they emerge during the development process.

There is no contradiction between this method of obtaining detailed requirements and the message that systems analysis is structured in stages which must be systematically traversed.

It is assumed that 4GT will be used in the writing of the application programs, whether or not the team choose to call it prototyping is irrelevant. The subject is covered in more detail in the final chapter.

Chapter 8

Designing and Using a Model of the Real World

PROCEDURE-DRIVEN AND DATA-DRIVEN ANALYSIS

By this point the project team has discovered why the application has been requested, and will have analysed the existing system (if it exists), and drawn up an activity chain and perhaps a DFD to decide what activities should be included in the new system.

The boundaries between the computer kernel and the rest of the application have been defined, and the information requirements of the new application and its users have been established, initially in broad-brush terms and later in more detail. The messages which should carry the information have also been identified, and their interaction has been scrutinized so that where possible they can be designed as turnaround documents to ease the operation of the proposed new application. The inputs and outputs to the total system and to the computer system have been allocated to the messages which will carry them. All this has been done by producing deliverable documentation.

All the progress so far has been made by purposely ignoring the computer kernel of the application. The only factors considered have been the business imperatives and the operational imperatives which must be designed into the application. Up to this point the analysis has been chiefly concerned with the required activities and messages of the new application. DP and MIS jargon calls this method 'functional analysis' or 'procedural analysis'.

As the project moves on to specify the computer kernel of the application the emphasis changes. It is machine requirements which become the new imperatives, and the emphasis of design is switched to highlight the organization of information within the computer. The *analysis* which until now has been primarily 'procedure-driven' changes to *design* which is 'data-driven'.

There is a debate in some DP circles about whether systems development should be procedure-driven or data-driven, and it illustrates a certain confusion between specifying the business problem and specifying the computer kernel.

The definition of the application in business terms must be a definition of procedures because the procedures literally are the application. Conversely, the definition of the computer kernel must be data-driven because the kernel literally is a collection of data which has to be defined before the operations or programs which process it can be defined.

It should be noted however that the analysis of procedures was linked very closely with the information requirements of the user and with the data provided to the computer. A clear distinction was made between the totality of information inputs and outputs and the messages which carry the information into and around the system. Procedures and information are complementary in administration so that it is hard to fix one without affecting the other. However, while the emphasis in the business study is on procedural analysis, the emphasis on designing the computer system is on data analysis.

THE APPLICATION MODEL OF THE REAL WORLD

The last point leads on to the design of a new application. The application should always process information about the real world where real people buy real products; in this sense it is a model of the real world. Information about the real world should be the focus of specification. The organization of the model is a mixture of real-world information and synthetic procedures for handling it. The procedures are fixed solely by business and operational requirements.

Accordingly, the aim in designing the new application should be to imitate what happens in the real world and not to imitate the old application procedures which may not be a true model. That is why the distinction was made in the business study between information and the messages which carry it.

This is a weakness in many development methodologies which proceed seamlessly from an analysis of an existing situation to the design of an application for mechanizing it. For instance the need for documentation to red-flag production orders might disappear if procedures for production schedules are changed so that computer-derived schedules automatically take account of potential overdues.

A further reason exists for a data-driven emphasis in the design of the computer kernel. Activities in the real world change about four or five times as quickly as the information which they manipulate. The same kind of information (on sales, stock, purchases) is needed in retailing applications as was needed a century ago, but the mechanics of selling, purchasing and procedures for controlling inventory and credit have all been revolutionized.

If an application is viewed as a collection of information which is systematically arranged or structured, the application can be extended by writing additional programs. But if it is viewed as a collection of interlocking procedures, changes are more frequent and are so intricate and messy that change becomes impossible. This is one reason that many old computer systems are inflexible. The design procedures were incorrect.

THE DATA MODEL

In the 1960s a series of writers made the distinction between the real world and the computer model. They preached that the computer model should be constructed to contain information about the real world.

The real world is made up of entities, according to this theory, and the computer files, one file per real world entity, should contain information

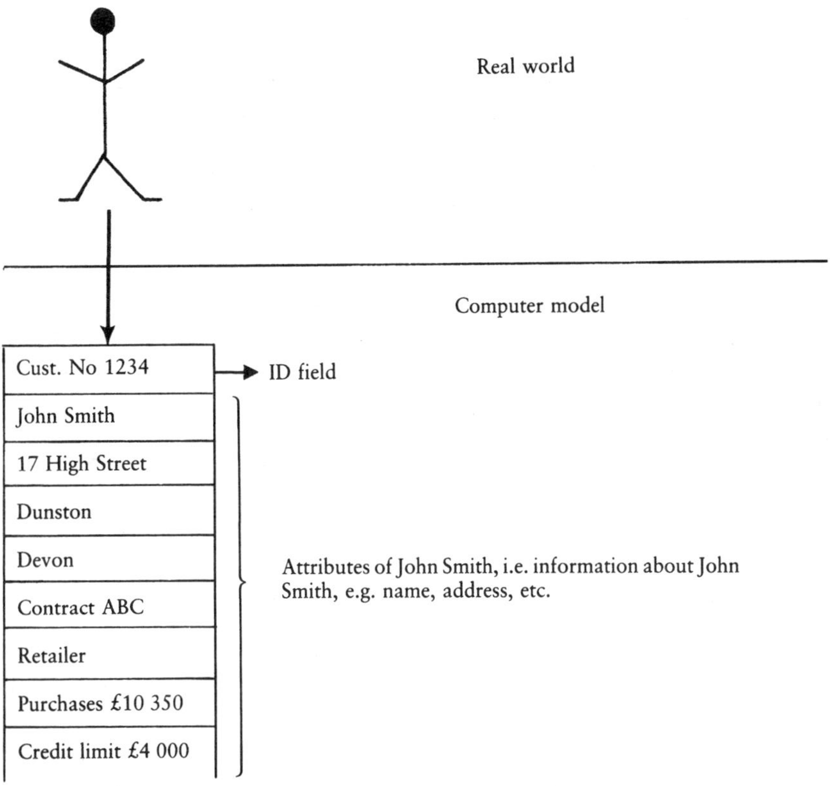

Fig. 8.1 John Smith (customer no. 1234) with his attributes.

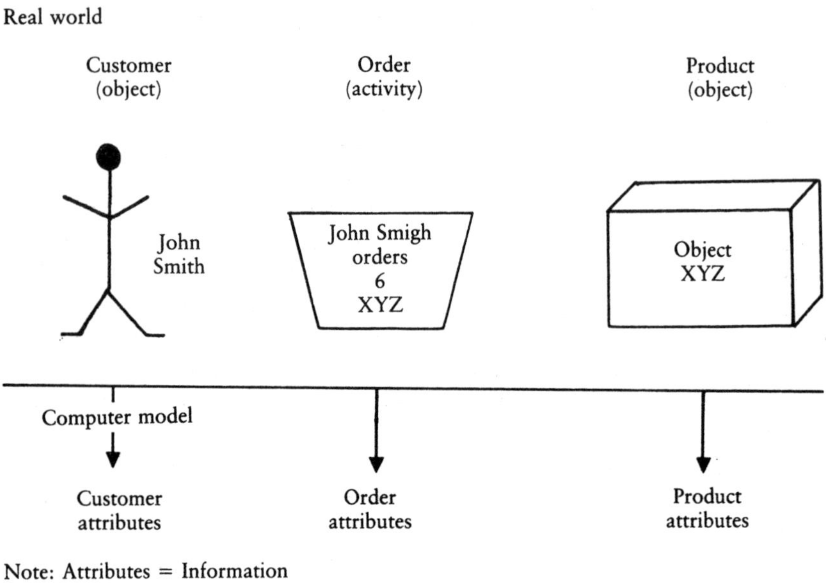

Fig. 8.2 A customer, a product and a sale.

about these entities. The coded information in the files is given the name of 'attributes', and confusingly the identification field for the file is also called an entity. A customer entity is therefore referred to confusingly in DP both as a real person who walks around in the real world and as a data field containing an identification key to the customer records.

Figure 8.1 shows an example of the real world with John Smith, customer no 1234, with his attributes. The concept of the data model is correct as far as it goes, but for purposes of systems design it has major shortcomings. It makes no distinction between an activity and the objects which become associated to perform the activity.

An entity in DP is either an object, or a concept, or an activity in the real world, as well as being an ID field in a computer file. A consequence of this failing to distinguish between objects and activities has already been met when obtaining information requirements from managers.

Figure 8.2 shows a customer, a product and a sale. The manager is able to say what he will need to know about customers and classes of customer, and about sales items and classes of items. He is never able to say what he will need to know in the future about the sales activity. Other shortcomings will emerge as the process of systems design is followed.

LOGICAL AND PHYSICAL MODELS

Another distinction has been developed in recent years between the way in which the application designer regards information and the way in which the information is arranged so that it can be efficiently manipulated by the computer. For instance customer file records are viewed by the designer as being logically next to each other in the same file. However, the computer operating system or DB software system may be designed to mix customer records physically with other sorts of records if it is operationally desirable.

There is a third level of information organization, i.e. the way in which the information appears to be arranged to the user of the system. There are usually many 'user-views' in a single application. The user will view the order book application as a collection of orders, enquiries, pick lists, inventory records, statistics, etc.

The logical schema for data provides maximum potential flexibility in using the data. Once data exists in a logical file it can be pasted into as many different user-views as required.

In the order-book application, the customer address may be needed on an invoice, a dispatch note, and an order, etc., all of which are user views of the information. As far as the user of the system is concerned the address must exist in all three places. In fact at the logical level there is only one address in a single location which is made available on demand for each activity or user view.

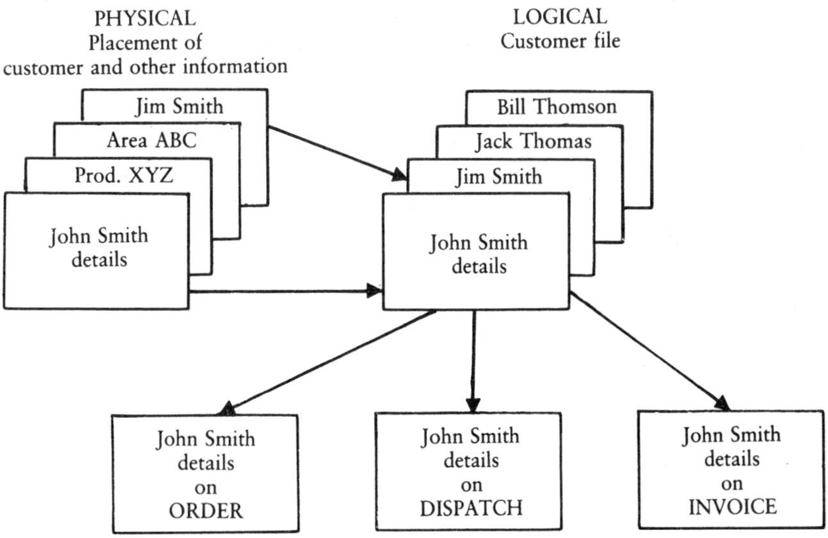

Fig. 8.3 Organizations of information in a computer kernel.

There are three separate organizations of information in any computer kernel. They are as follows:

1. The physical as viewed by the computer.
2. The logical as viewed by the systems designer.
3. The user view as viewed by the user.

Figure 8.3 shows these three organizations of information.

The first data organization to be defined is the logical organization which should be laid out in such a way that both the physical and the user-views may be obtained automatically from it.

Fortunately the latest database management systems (DBMS) contain software to derive physical organizations (or 'schema') automatically from logical schema, and they also provide facilities for user-views to be derived automatically. The structuring of information into the three-level model of a DB is covered in the next chapter.

BUILDING THE DATA DIRECTORY

Before the information in the object listings and the message content lists can be organized or structured into a logical schema it should be checked for completeness and consistency. At this stage therefore the team has to establish a data dictionary or data directory (DD) for the application. A DD among other functions should cross-reference all data elements to their points of origin and to the messages on which they appear. It lists all the data elements in the application, the roles that they play in the application, and the way in which they are logically grouped into files.

Most modern 4GT systems incorporate a DD and where it exists it should be used as far as possible. There are also DDs which are associated with CASE tools and they too can be used. There is an enormous amount of cross-checking of documentation involved in the design of the computer system, and the use of a mechanized DD increases analyst productivity probably by an order of magnitude in this stage of systems development.

Most of the data to establish a DD already exists in object listings and message content lists, and it makes sense, if it is technically possible, to set up the DD right at the beginning of the project and to use it to produce object listings and message content lists in the business study via standard print-outs. The computer can then keep them updated as more information is entered. If this was not possible because of shortcomings in the DD, the data already obtained about the information base of the application should be put into the DD at this point in the design process, or associated with it by addenda.

In prototyped applications, the computer programs eventually may be generated directly from entries into the software DD, which are often and mistakenly made in a slapdash fashion to produce a set of programs quickly. But the DD must be tightly controlled. It is the key to successful prototyping. One can be informal, one can never be slapdash.

A major reason for using a DD, regardless of whether 4GT is being used to build the system, is that it serves as the base for the production of systems documentation.

An application must possess documentation to allow the application logic to be scrutinized and analysed prior to handover. 4GT does much of this checking automatically. No system can be accepted as operational without this scrutiny, as the fact that it has worked satsifactory until now is no guarantee that it will work satisfactorily tomorrow. Moreover without this documentation it will be very difficult to maintain and extend the application. The core of this documentation can be derived from the DD, which is a précis of what is the data and how it will be used within the computer programs.

A program is sensitive to the data it processes and it must handle all possible combinations of data. It is theoretically impossible to physically check that this is the case, so documentation must be provided to provide a basis for logical analysis of the application during development.

Within the DD, data fields are arranged in records, and records are arranged in files. Because the first design operation is to organize the data logically, the records under discussion are logical records.

In order to help in checking the consistency of the eventual design with the business study, the data entries in the DD are split into simple categories.

1. Object files: A list of all the logical files about objects.
2. Object attributes: A list of data elements for each object file.
3. Activity files: A list of all the logical files about activities.
4. Activity attributes: A list of data elements for each activity file.

These correspond very broadly to entity definitions and user-view definitions. An activity is always depicted via a user-view but not all user-views represent activities. *Note*: The split between object files and activity files will not be handled by the DD on the computer. The distinction should be maintained as a note on all data documentation.

The simplest way of relating the data structure to the business study is to examine *in conjunction with the end-user* all the fields on each message in the message content list, and to place them into the categories above. The object files and attributes should also correspond with the object listings. A further check, which does not take much time, should be made that this is so. This is a matching operation – matching design back to the original business and operational specifications.

The order enquiry message, for instance, contains information about the customer and the product. Individual fields are given a systems name and are entered onto the object attributes file lists for customer and product respectively, whence they are cross-referenced back to the order enquiry message. Once again this might not be handled by the DD and might have to be inserted as an addendum.

Most of the information contained on the order enquiry will in fact be listed as belonging on the object files.

Most of the attributes of an activity are the attributes of the objects which are associated to carry it out. However, there will be other information which is always recorded for each individual activity. In the order book application this is restricted to the following:

1. Sales point or sales rep. ID.
2. Customer ID.
3. Product ID.
4. Message ID.
5. Quantity involved in the transaction.
6. Unit of measure (may be implicit).
7. Time of transaction (perhaps).
8. Date of transaction.

All activity records have this information skeleton, i.e. pointers to the two or three objects that are associated in the activity, an activity ID, a quantification of the activity and a date stamp of some sort.

There is a significant difference between the fields associated with objects and those associated with activities. With activities it is likely that there will be repeating groups of fields. An order or an enquiry usually contains more than one item. It is constructed of an order header and repeating groups of data elements for each order item (see Fig. 8.4).

On the order enquiry there may be information fields which can be deduced from other information fields (e.g. the value of the eventual order). These derived fields should be indicated because the designer may want them generated each time that they are required by the system rather than calculating and storing them for future use. They are called 'virtual' fields.

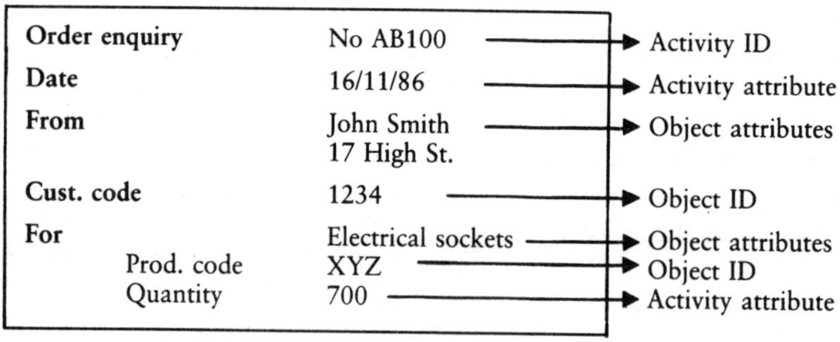

All fields can be checked against object lists and message content lists

Fig. 8.4 Activity DD entries from an order enquiry

An example of a virtual field is the 'value' of an order obtained by multiplying 'quantity' by 'unit price'.

OBJECT CLASSES

Once the message content lists and their associated object listings have been transferred into the DD object and activity files, the required attributes will be known for all the individual objects and activities in the application. At some later stage it might be necessary to reconsider the messages which carry this information. But before that happens there are other kinds of objects to be considered.

A customer may be categorized in various ways. Some of the more common classifications are as follows:

1. Sales area or region.
2. Industry classification.
3. Usual method of delivery.
4. Type of customer (retail chain, wholesaler, wholesaler chain, buying group, etc.).

All these classifications become conceptual or derived objects in their own right, as they all have attributes. The minimum attribute for a 'method of delivery' derived object will be a line of text printed onto every dispatch note to every customer in that delivery category. There may also be standard costs, cumulative totals, planned utilization etc.

The next step is to consider the data for derived or conceptual or classes of objects, again using the object listings, and again relating each data field to an entry message (see Fig. 8.5).

The data field indicating the type of customer on the individual customer record, i.e. the primary object record, should be marked to show that it points to a conceptual entity. It is indicative data which because it points to another file of which it is the key becomes '*key indicative data*'. It is also known as an entity ID.

Other roles for data fields are either to carry quantitative data, or to carry unit of measure data, or to carry non-key indicative data, (an example for a product might be whether it is a high, or medium, or low turnover item). Wherever indicative data points to derived objects (a rule of thumb is when there is more than one attribute field associated with the conceptual class) it becomes *key indicative data.*

Most of the data has now been copied into the DD and checked against the original object listings and message content lists where it originated in the development documentation. Any expansion from the original documents will come about as management requirements are discussed with the end-users and compared with existing derived objects. Note again that the DD is

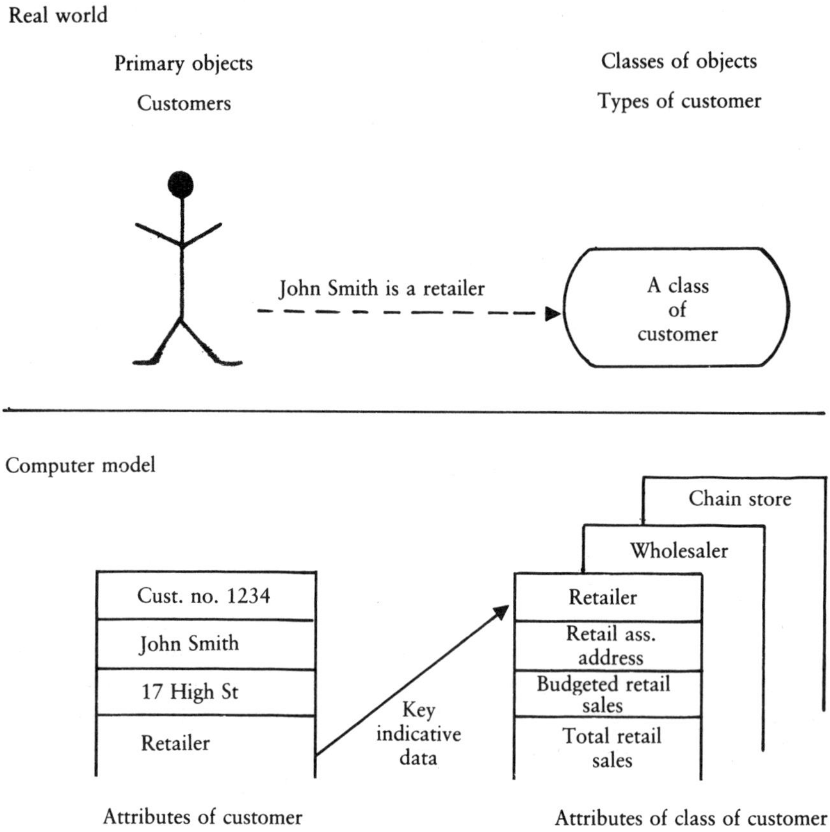

Fig. 8.5 A derived object.

related back systematically to the original business and operational specification of the application.

Management initially provided broad-brush information requirements on the object listings and this was expanded to give more detail as the message content lists were developed. The opportunity is taken on setting up the DD to check that any additional detail that has arisen has been considered.

In most cases a classification code will already exist. Additional classifications (new key indicative data fields) will cause new object files to be raised and additional requirements will cause new attributes to be entered.

For instance it may have emerged that the sales area classification of customers needs extending to allocate customers to standard delivery routes. This may involve the creation of a new object class if many data elements are to be associated with this new grouping. Alternatively it may suffice merely to insert an additional attribute in the customer primary object record.

GIVEN ATTRIBUTES AND SYSTEM-GENERATED ATTRIBUTES

Information is either presented directly to the system or is generated as a result of processing messages. For instance the credit limit of a customer may be derived on a continuous basis from an analysis of his payment record (e.g. equal to what he has paid in the last three months).

Data fields should be marked on the DD as either given or system generated. The reason for distinguishing between static (given) and system generated fields is that most program modules are concerned preponderantly with one or the other. When it comes to designing for performance, it might be useful to group like fields together at the physical level.

UPDATING THE DD

The final stage in setting up the DD is to note any data fields which are not cross-referenced between object listings, message content lists and the object and activity files in the DD. Such fields might have originated on the original object listings and slipped through the mesh, or they might be additions that were not associated with a message during the business study. By noting in the DD the message, or messages, or origin of the field one checks automatically that all information can be eventually entered into the working system.

A formal check should be made that every data field is contained on messages that will: (a) enter it into the system; (b) process it or interrogate it; (c) modify it, and (d) delete it. This is the activity carried out by entity life history diagrams in traditional methodologies and is a good example of how 4GT can sometimes cut the amount of development documentation.

Note that this procedure tends to produce a default design that any object information which can be carried on an activity message will create object records the first time that it occurs, i.e. a customer record is established by processing his initial order. This is because if information happens to exist on an activity message there is a logically no need to provide a duplicate message to enter it into the system.

The default design should be checked against the labour balance and message flow charts to see if it is desirable. The aim should be to put as much information as is possible and ergonomic onto a single document, but at the same time to allow partial processings of the document to allow procedures to be simplified. This raises points of interface design which are outside the scope of this book.

The final action is to proceed through the object file lists in the DD to check that information which exists on more than one kind of message will always be available for all messages, e.g. if a customer record can be set up *only* from a customer order, how does one log an enquiry from a *potential* customer?

A different kind of common design error is to nominate output data for a message before the input message has made it available. This is avoided by using the activity chain document. Each output message on the computer box diagram should therefore be checked to see that its input is provided at a prior stage in the activity chain. Again the control documentation is that used to establish business and operational specifications.

A final check is made to ensure that virtual or derived fields in a transaction processing application do in fact reflect the real world which they purport to model. For instance if messages for receipts into stock are entered in batch into the order-book application from an application for production control, the stock-out messages produced by the transaction processing order-book application will sometimes be incorrect.

Most design methodologies do not check either the sequencing or the timing of messages and this is a major weakness.

Chapter 9

The Relational Model and Third Normal Form

RELATIONAL DATA

Note: There is nothing new in this chapter for the computer professional. It explains to managers and end-users the theory behind the construction of a database for a DP application – the key activity in specifying the computer kernel of an application. It is also of value to the end-user engaged in the design of a departmental system under the general guidance of a DP professional.

The twin concepts of relational databases (RDBs) and arranging data in third normal form (3NF) are so closely connected that it is difficult to talk of one without the other. A file in an RDB looks like an enormous matrix with each record taking up one line. Relational data is arranged in such a way that some of it can be abstracted from particular records in the file (i.e. lines in the matrix) and connected to data similarly obtained from other relational files or tables in order to construct a new file or matrix.

The records are 'selected' from a matrix, the required data elements from the records are 'projected' into a working file or new matrix, and finally they are 'joined' to other data elements which have also been selected and projected. Each of these operations usually only requires a single program statement. A relational structure provides an ideal logical schema for a nontrivial DP application.

The operations of selecting, projecting and joining might produce nonsense files unless the original data was arranged with them in mind. For instance each relational file must only contain one kind of record, and there must be matching entities in the two files to be merged. If data from an overdue order file is to be joined with data obtained from a customer file then every order in the overdue file must be for an overdue order, and must correspond to a customer on the customer file.

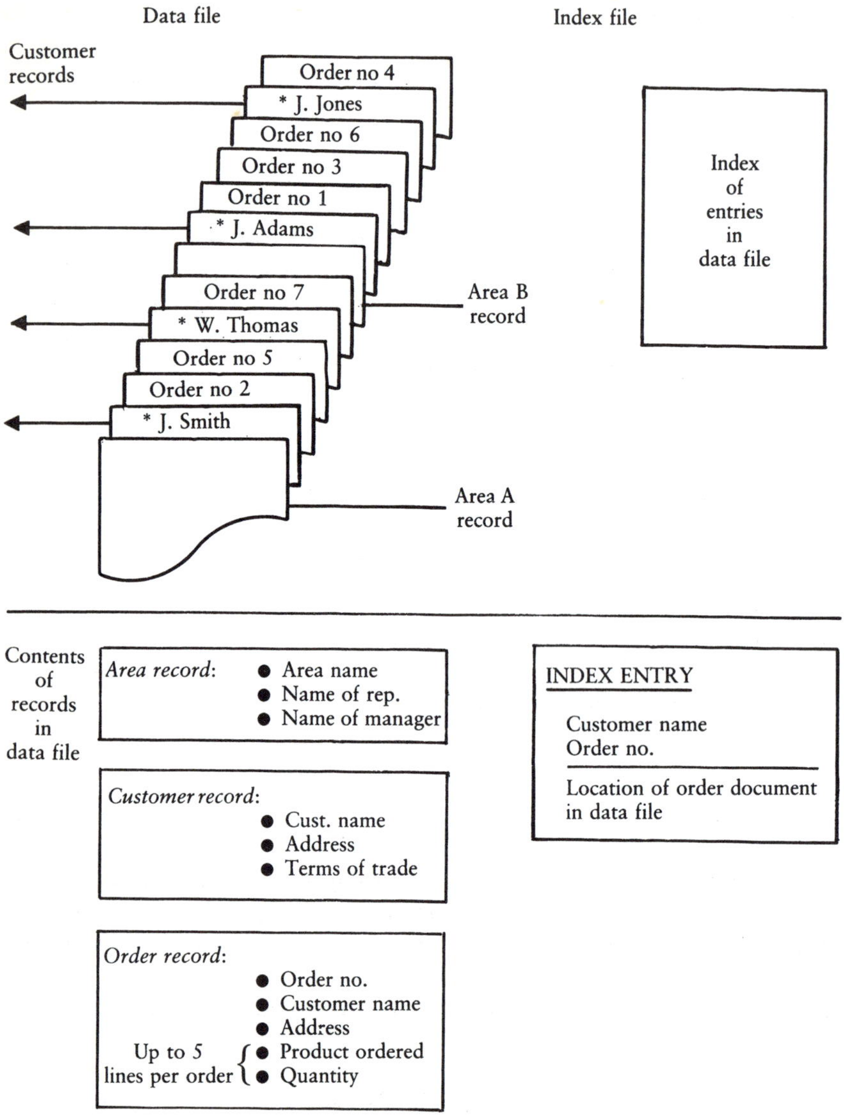

Fig. 9.1 Card index grouped by area.

THIRD NORMAL FORM DATA (3NF)

Relational organization of data only pays off properly if the data is in 3NF, and conversely if data is in 3NF it fits naturally into a relationship organization. 3NF is shorthand for saying that data elements must only be written once in a logical DB and retrieved by cross-reference when needed for processing or for output. Much of the software for relational data organization has been incorporated into relational database packages, but it is not necessary to have an RDB before data can be organized in a relational manner.

The best way of stripping the jargon from an explanation of 3NF and RDBs is to explain how a relational organization could be constructed using 3NF data on a manual filing system. It would never be implemented in practice because a computer is needed to deal with the extensive cross-referencing, but it shows how the computer, when used as an electronic filing cabinet, provides facilities which a manual system lacks.

THE RELATIONAL FILING SYSTEM

In a manual system for processing orders and related activities there very probably would exist a file containing information about each of the customers. It would be sequenced in alphabetical order and would contain initially merely the name and address of the customer. If a new customer, Jim Smith, is accepted his record would be slotted into the file just in front of John Smith.

In order to answer queries on the status of customer orders, e.g. released to the warehouse, dispatched, partially delivered, etc., the manual system places additional documents into the customer file so that all the orders for John Smith would be placed physically into the file immediately behind the record that contained his name and address. This is now a 'header and trailer' file with two different kinds of records: a customer 'header' record which applies to all the order 'trailer' records which follow it.

Assume that it was discovered that planning the delivery of completed orders would be eased if the records were arranged into groups for each delivery area. Accordingly a new kind of 'header' record was added to the file to carry information about deliveries in that area. The customer records and the order records were then rearranged behind the new area header card (see Fig. 9.1).

At this stage there came the first problem: it was no longer possible to find a customer record simply by seeking the name of the customer, it could be in any of the batches behind any of the area cards. To overcome the problem a card index was set up. The index sequenced in alphabetical order carried the information that the record for John Smith is say the fifth customer record in area two.

HIERARCHICAL DATABASES

The file is now hierarchical. It is organized with an area record containing information which refers to all the customers in the area, a customer master record containing information which refers to all orders placed by that

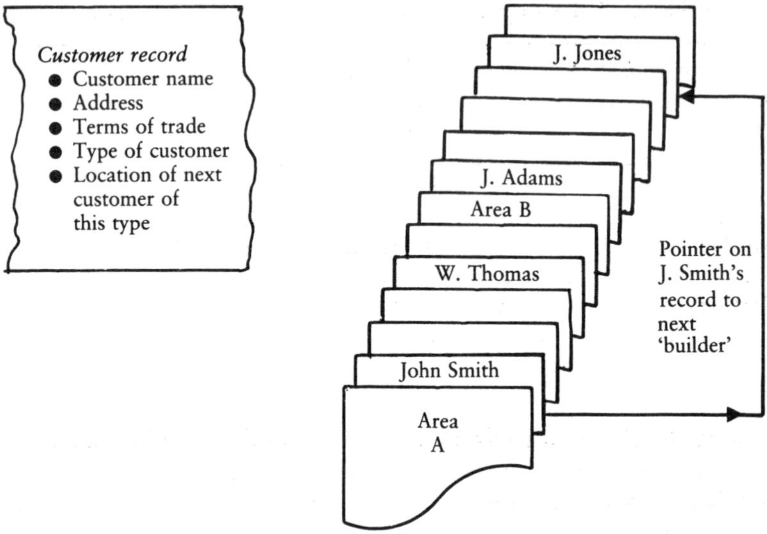

Fig. 9.2 Card index with internal pointers.

customer, and a series of order records. The physical location of the records corresponds to the logical hierarchy and their location is indicated in a separate card index. It is a very efficient way of storing and retrieving information about areas and customers and it is basically the way that hierarchical databases are organized on a computer.

A snag arises when the system becomes still more complex. Assume that the marketing people want information about orders from specific industries; construction, engineering, mining etc. Accordingly they request that the file be reorganized so that the header record becomes not a card for the area but a card for the industry. The same hierarchical system can be set up, but it would mean the destruction of the existing area hierarchy. John Smith at 6 High Street might be a builder and his neighbour at 8 High Street might be a garage, so that the two records would be in the same area group but in different industry groups.

One solution is to expand the information on the actual data record. It is possible to leave the records in the original area sequence but to write onto each customer record the location of the next customer record in industry sequence. Jim Smith's record in one area would indicate that the record for John Smith (the next builder) was in another area section of the file. Jim Smith's record 'points' to John Smith's record (see Fig. 9.2).

Thus besides the physical hierarchy of the actual records in the filing cabinets with a card index to ease retrieval, there is another logical hierarchy which can be established by following pointers. Alternatively the pointers can be abstracted from the records and placed into an additional secondary index (see Table 9.1).

Table 9.1 Index to process customers for areas within industries

Type of customer	Area	Customer name	Location
Builder	A	J. Smith	
Builder	B	J. Jones	
Retailer	A	W. Thomas	
Retailer	B	J. Adams	

A hierarchical data organization assumes that data will be used for one purpose and arranges its hierarchies accordingly. Secondary indexes may be established (as above) but the assumption is made that different data usages will match the sequence of the keys in the different indexes that have been constructed. For efficient operation a separate index is needed for every logical hierarchy that will be processed by program logic.

The distinguishing logical characteristic of the hierarchical database is that each record can only have a single header record in each hierarchy, e.g. a customer belongs to a single area, but can be cross-referenced to a single industry.

NETWORK DATABASES

A network DB accepts that a record can exist in more than one place in the same hierarchy, it can have more than one parent. John Smith for example could be both a builder and a civil engineer and exist in two different places in the industry hierarchy. A record can have more than one parent and inherits the characteristics of the relevant parent when processed in that parent's hierarchy. A child belongs to the families of both its father and its mother. A record in a network DB can point to a variety of other records as the next higher level in the same hierarchy.

There are again two ways of dealing with it in the manual file example. One can write pointers onto the actual data records, or an extra index can be constructed. In the latter case the key to an individual record can be entered into the index as many times as necessary, e.g. John Smith would be found both in the builders' and the engineers' sections.

The difference between hierarchical and network DBs is shown in Fig. 9.3.

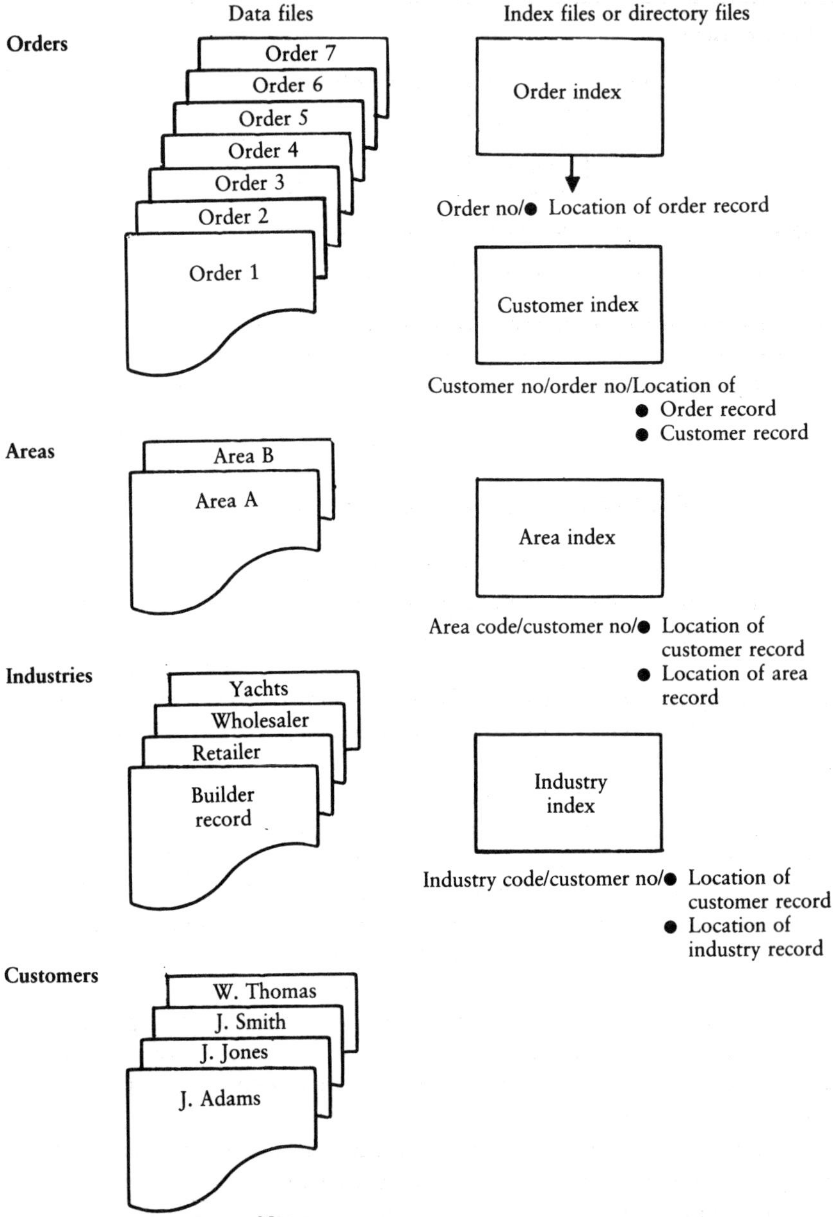

Notes

1. The logical hierarchies are (a) built into the indexes as above or (b) indicated by pointers written onto the datafile records (i.e. loops).
2. In a network DB a customer could belong to two industries at the same time.

Fig. 9.3 Card index example.

Network indexes

Key 1 Truck no./Area/Industry/Customer/Dispatch no.

Difference in index sequences

Key 2 Truck no./Industry/Area/Customer/Dispatch no.

Which index to structure the derived information regarding trucks? Key 1 or Key 2 or both?

Note: Fixed hierarchies of information are effectively coded into the indexes.

Fig. 9.4 Indexes in a network (also known as plexed) databases.

DRAWBACKS OF FIXED HIERARCHIES

Hierarchical and network DBs are inflexible because data will be retrieved efficiently only by following the sequence of the hierarchies as set out in the index. If you want to find John Smith's record you must know that he is either a builder or an engineer. If you were searching for a friend's telephone number it is possible to find it quickly only if you know in which directory it is to be found. However, if you only know the county of residence it will take several minutes to peruse all the directories where he might conceivably be listed.

There are further drawbacks. Suppose that the filing system has to be extended to incorporate dispatch notes as well as orders. In this case we wish to file the notes behind a header record for the truck which carried it. But the trucks are not associated either with an area or with an industry, so that a third independent hierarchy is needed, but what exactly should it be?

Should it be a simple two-level hierarchy of truck header record and dispatch note trailer? Or is the hierarchy to ascend from dispatches, to customers, to areas, to industries, to trucks? Or should it proceed from dispatches, to industries, to areas, to trucks? Or should there be some other permutation (see Fig. 9.4)?

If the index does not match the way in which information is retrieved the filing and finding of individual records will become complicated and very time consuming.

RELATIONAL DATABASES (RDBs)

The RDB aims to solve these problems. It says briefly that all the information which is exclusive to a customer shall be carried on the customer record, and

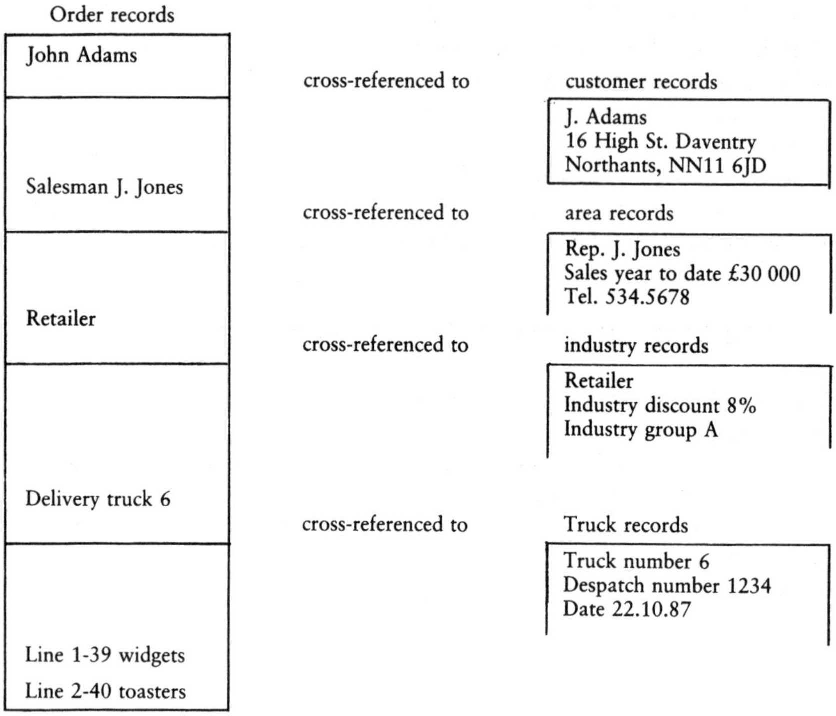

Fig. 9.5 Relational records in a card index.

nothing else shall be carried on the customer record. Likewise for trucks, areas, and industries.

In an RDB, a logical record contains the information, the whole information, and nothing but the information, that pertains to the owner (entity) of the record.

Individual information fields can then be nominated, e.g. the standard delivery charge of area six. This automatically avoids the problems of hierarchical and network DBs.

The first stage of establishing the RDB manual system is to set up a file for orders, another for customers, another for areas, another for industries, and another for trucks. A file is set up for each entity (which in standard DP theory can be either an object or an activity). Then each order is cross-referenced to its customer, its area, and its truck for delivery. The logical record for the order contains indicative key fields which point to the associated customer, area, and truck records (see Fig. 9.5).

The idea of header records and trailer records in the same logical file is absolutely forbidden by the RDB. There are no hierarchical levels of information in RDB logical files.

Another required characteristic of an RDB is that all the records in a file shall have an identical format. There cannot be one record which gives a delivery address at the top of the form and another which writes a telephone contact in the same space. The matrix of data is the same as a matrix in algebra, every entry in a column must refer to the same attribute class. The range of entries in the matrix column incidentally is called a domain.

The final requirement of an RDB is that information on it is written in 3NF; this is a logical consequence of the other two requirements.

NORMALIZING DATA AND THIRD NORMAL FORM

The logic behind 3NF is quite complex but luckily it is a simple concept to grasp intuitively.

Files are in 3NF when all information is written once only in the entire set of files regardless of the number and variety of records in which it is utilized (in theory this is not 3NF but 5NF but no one ever bothers very much about 4NF and 5NF).

For instance a customer address should be written once on the customer record, it is not written onto every single order record (see Fig. 9.6).

The procedures for turning data into 3NF are fairly simple. They are as follows:

1. List all the data fields in an unnormalized record.
   D1.
2. Take out all the repeating groups of fields and put them into a separate record.
   D2.
3. Take out any data fields which are not unique to the entity (i.e. can also apply to a *different kind* of record) and place into further records.
   D3.
4. Take out any data fields which are associated with normal attribute fields (e.g. can also apply to other records in the *same file*) and place into further records.
   D4. (anagraphic).

Figure 9.7 illustrates these procedures.

However, although normalizing data is simple it is tedious because there is so much of it. In the order-book example there might be typically 600–900 data fields to be normalized. There are computer packages which help to normalize data and it is recommended that one be used.

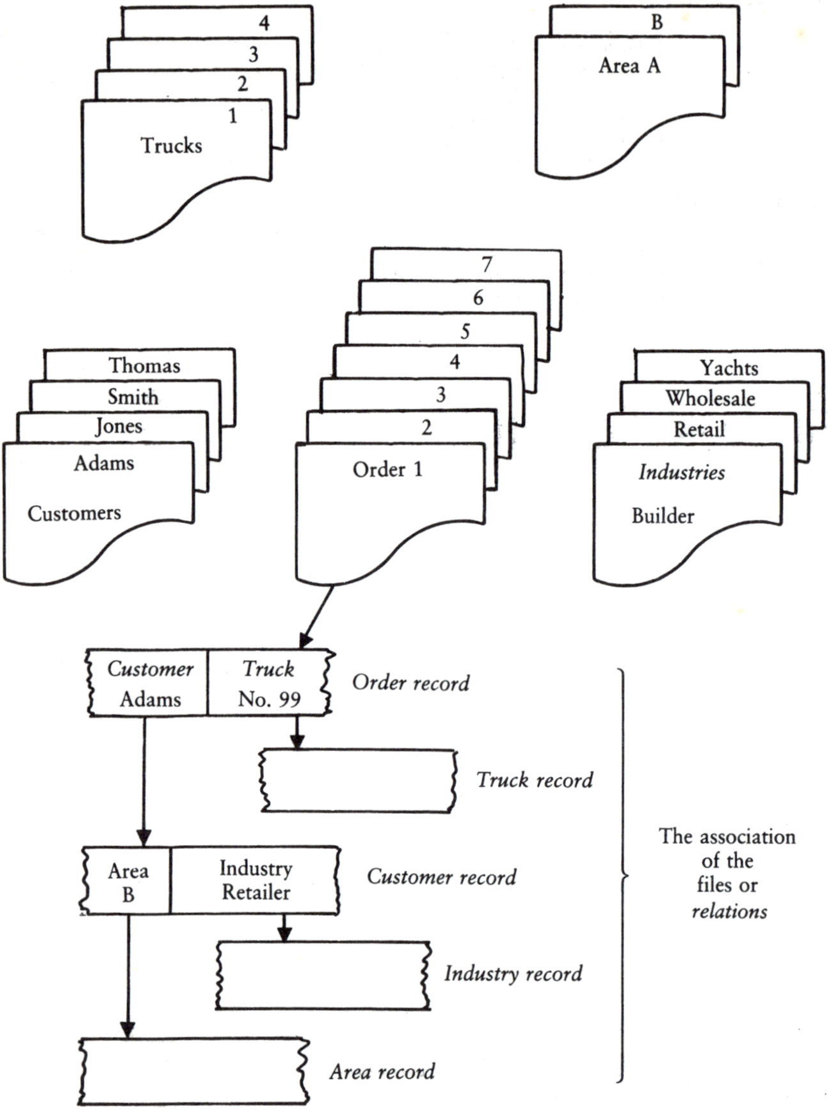

Indexes are:

	Key	*Location of*
1.	Order	Customer record
2.	Order	Truck record
3.	Customer	Area record
4.	Customer	Industry record

Fig. 9.6 3NF – customer, industry and area.

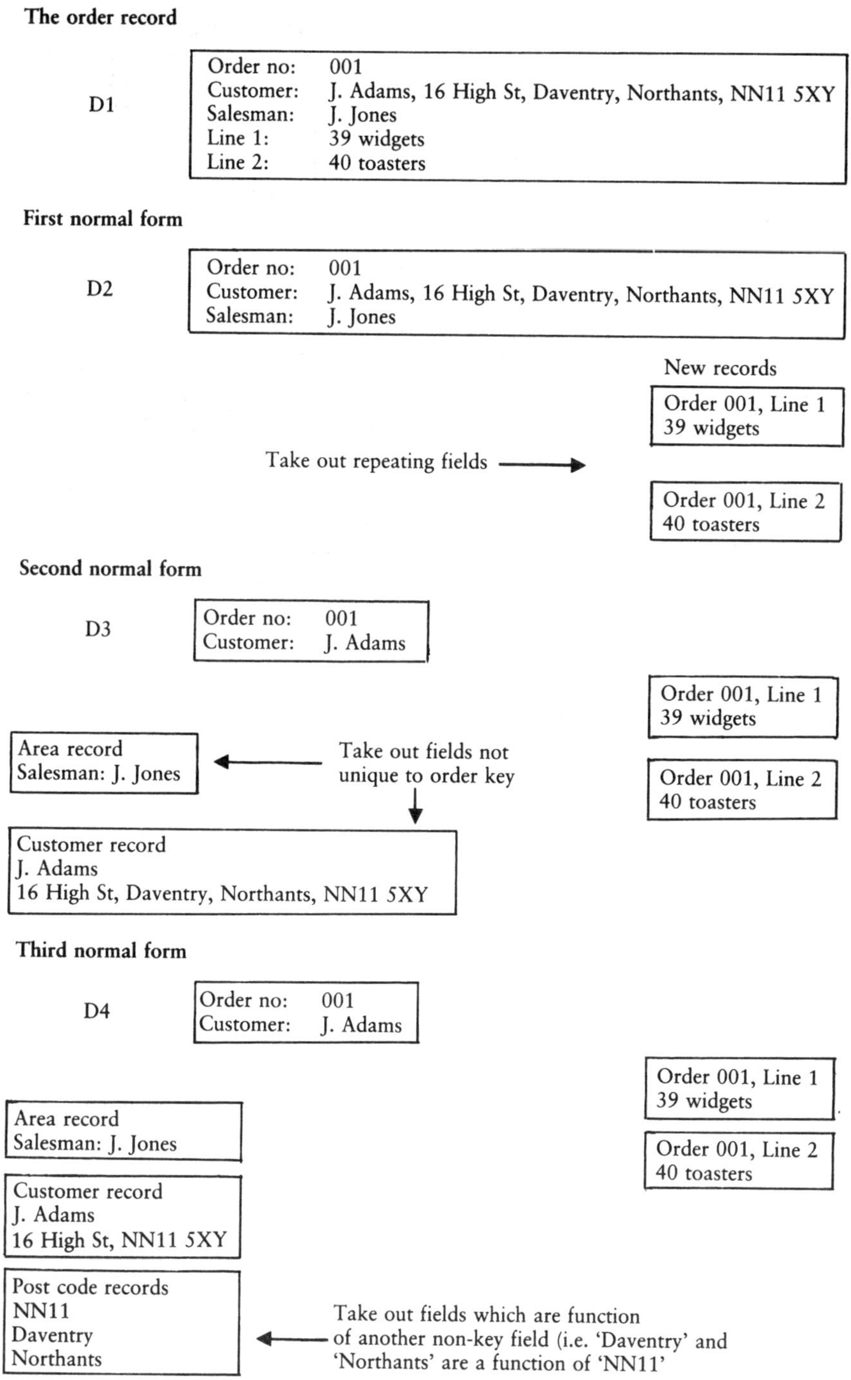

Fig. 9.7 Third normal form data.

THE ADVANTAGES OF RDB AND 3NF

1. An RDB saves space.
   Assuming that two hundred customers live in Northamptonshire or 'Northants' and the address is written in 3NF. The saving is 199 × 9 characters (on writing 'Northants' once instead of 200 times) less 200 pointers in the customer records (say 4 digits) to the 'Northants' record.
2. It makes DB maintenance easier.
   Suppose that industry details change or need expanding – perhaps a new classification for VAT purposes. The change can be made on a single record in 3NF, and not on hundreds of individual customer records.
3. The system is more flexible.
   If it is decided later to introduce more activities, such as the dispatch of samples or special offers, all the information in the RDB can be easily assembled to support the new activities. There is no need to construct and optimize new hierarchies and to laboriously try to guess the optimal sequence of additional indexes.

It is therefore desirable to use relational organization of 3NF data within the computer kernel of the application, even if the computer will eventually use either individual file handlers, hierarchical or network DBs to manipulate the records.

Chapter 10

Designing the Logical DB

The distinction which has been established between primary objects, classes of objects and activities helps in the specification of the logical DB schema. Nothing in this chapter is incompatible with standard development methodologies but there are some suggestions as to how they should be sequenced. There is also an additional key design activity of commonizing files, which is usually inadequately treated in standard methodologies. This is therefore a chapter for the DP professional as well as for the manager and end-user.

Designing a logical schema is a process of refining the DD entries, which were based exclusively on business and operational requirements, to develop them into a set of similar files which are oriented to the requirements of the computer kernel. In DP these are ease of maintenance and ease of extension, so that eventually the application can be integrated with neighbouring applications. The procedures so far have merely checked the consistency and completeness of the entries in the DD.

The decision to use the relational model as the basis for the logical design of the DB has the incidental advantage of connecting the design of the DB with the documentation that has already been produced. The design is based on the use of object listings, message content listings, object files in the DD and activity files in the DD, which have already been produced. By utilizing this documentation a check is introduced on the definition of the logical files and the design of the DB is associated with the business study. In theory, DB logical design can operate exclusively on the DD entries but it is generally worth checking the consistency of development documentation by using all of it for major activities. If the manager has added something to the object listings after the DD has been set up this checks that it has been fully considered.

This chapter describes a design procedure for a relational data structure, as is normal today. The following chapter shows how this approach may be modified to produce a data structure which is usually more efficient in operation.

LOGICAL FILES FOR PRIMARY OBJECTS

The first set of files to be identified belong to simple individual objects in the real world. In the order-book application the files may be as follows:

1. Customer.
2. Product.
3. Warehouse/dispatch point.
4. Salesman.
5. Point of sale.

As all orders in the example are processed centrally and dispatched centrally there is no imperative need for a point of sale file or for a warehouse file. Any information about either of them can be safely stored as a table in an 'omnibus table file' which operates outside the DB.

Once the primary objects have been chosen, the next step is to normalize the data associated with them on the object listings, the message content lists and the DD files. The procedures described in the previous chapter produce the following normalized data for the salesman:

1. Salesman ID code.
2. Salesman name.
3. Area code.
4. Total sales value YTD.
5. Manager ID.
6. Base office ID.

This could provide import/export facilities to a future application in marketing.

The procedure for primary objects is as follows:

1. List the primary objects.
2. Scrutinize their object listings and DD files and check that all data fields belong to the primary object.
3. Check that codes exist for all key indicative data fields required for derived objects.
4. Scrutinize the message content listings and check that all ID fields point to a primary object logical file.
5. Check again that every field in the primary object file is contained on a message and indicate its message(s) of origin.
6. Normalize the data in the primary object files to develop any related files for data which belongs to more than a single object. This should automatically indicate files for derived objects (e.g. sales area) as well as the files which emerge to eliminate data redundancy (e.g. the post-code file).

LOGICAL FILES FOR DERIVED OBJECTS

Once the attributes of the primary objects have been gathered and normalized, the next group of logical files to be tackled is concerned with derived objects, e.g. sales areas, class of customer, industry code, class of product, etc.

The first procedure is to check each of the object listings and DD files for derived objects against the logical files that should have emerged from normalizing the data of the individual objects. The need for an industry logical file for instance should already have been established by normalizing the data associated with the primary object, i.e. the customer. Additionally the contents of the file should have been partly defined by the normalization process. Check each field which has been so defined against the message content list and indicate its message of origin.

The derived object listings, the message content listings and the DD files are now scrutinized and the procedure is repeated as for primary objects, relating

Customer (primary object)	
Customer ID	78912
Name	T. Jones
Address	50 Main St, Anytown
Credit limit	500
Industry code	123
Industry (1st level derived object)	
Industry ID	123
Industry ID	Ferrous casting
Budgeted sales value	2 000 000
Sales year to date	587 234
% budget	87
Industry grouping	12
Industry grouping (2nd level derived object)	
Grouping ID	12
Grouping name	Heavy manufacturing
Budgeted sales	14 000 500
Sales year to date	3 000 000
% budget	57
Industry class	5
Industry class (3rd level derived object)	
Class ID	5
Class name	Manufacturing industry
Notes on class	Excludes extraction industries
etc.	

Fig. 10.1 Industry data structure.

all fields to the message of origin. This is a check that the DD files have been kept updated during the development process.

Finally the data is normalized. It is at this stage that the number of logical files may multiply. In order to keep things simple for management and the user, the derived objects were all treated as features of a primary object. For instance the country might be split into a dozen sales areas which are grouped into three sales regions. Both the area and the region on the object listings are treated as derived objects pertaining to the customer. When the data is normalized, however, it will be apparent that the region is in fact a derived object pertaining to sales area. There are hierarchies of derived objects and each one should have a separate logical file (see Fig. 10.1).

In real life, an application of the size being discussed might contain 40 to 70 logical files chiefly because of this multiplication of derived object files.

LOGICAL FILES FOR ACTIVITIES

The final stage of setting up the core of the logical RDB is to deal with activity logical files. Traditional data modelling sets up entities for each of the activities which result in a document processed by the system. These include enquiries, orders, warehouse make-orders, pick lists, dispatch notes, bills of lading, receipt notes, balance orders, etc. The procedure of checking against existing documentation is exactly as for objects, except that there are no object listings to be considered at this stage. The relationship between object listings and message content lists was established when the DD was created.

However, a further step for activities is to check that everything about the transaction is contained in the associated object files (i.e. customer and product) except:

(a) The quantity.
(b) The unit of measure.
(c) The time.
(d) The place.
(e) The date.
(f) The transaction ID.

Not all these fields exist for all activities, but the above list includes all possible fields pertaining to an activity which should not have already been processed. There is nothing about a transaction except the above fields which is unique to itself and anything else must be contained in the object logical files.

COMMONIZING THE LOGICAL FILES

In all design one objective is to 'simplify and add lightness', and the same is true of systems design. Generally speaking the fewer programs and files and record types that can be designed into an application the more elegantly the application reflects the requirements of user and management. Having established logical files of normalized data for all the objects and activities the next step is to see if they can somehow be reduced.

There is no way of reducing the different kinds of individual objects in the system, and therefore there is no way of reducing the logical files which depict them. In an RDB there is no way of reducing the different kinds of classes of objects as this entails the organization of data in non-3NF (e.g. regional information would need to be duplicated on every area record).

Conversely it is usually easy to reduce the number of logical files which depict activities.

Conceptually an activity always changes the status of the record depicting the prior activity in the activity chain; except for the very first activity in the chain which establishes the record.

In the case of the order-book system, the basic activity is obviously an order, and the order record may be established either by an order enquiry or by a customer order. Thereafter every activity changes the status of the order record. It is much simpler to view all these subsequent activities as a variation of a basic message which says simply 'change the status of the order record'. This axiom of application design is rather masked in traditional design methodologies by the failure to distinguish between activities and objects.

Traditional methodologies do sometimes reach the same conclusion on a bottom-up analysis of required program modules. It is much easier to proceed top-down on the basis of the above axiom about activity messages. In customer accounting a sale will create an invoice record. Every subsequent activity, e.g. statement, overdue letter, payment, credit and debit note, etc., changes the status of that record. In material control a purchase order raises a record and each subsequent activity of delivery or part delivery, goods onward inspection etc., changes the status of the record.

In the order-book application there is an order record modified by all the other transactions. Instead of depicting a completed pick list document as a separate kind of record in a pick list file, it is regarded as an order modification activity. Within the pick list record a single ID field distinguishes it from all other 'modification records' such as dispatch notes, balance orders, goods received notes, etc. They will be found to follow the format below:

1. Transaction ID (identifies the modifying activity).
2. Document No.

3. Order No. ID.
4. Line No. ID.
5. Update Qty.
6. UOM (implicit in most cases).
7. Product ID.
8. Customer ID.
9. Date.

The commonizing of messages is so important in systems design that all the activity messages which cannot be commonized by consideration of the activity chain should be compared with all other messages to see if there is a duplication of data. Where a majority of the data in different messages matches, there is scope for commonizing the messages and the logical files which support them.

For instance on build-up to gross pay in a complex payroll, a job ticket and a sick note might possibly be commonized (they both can instruct the computer to pay a certain number of hours at a particular rate). Data comparison shows an overlap even though the two messages are never associated outside the computer in an activity chain. The aim is to find where messages are variations of an activity and then to design a general message to handle many different activities.

STATISTICAL INFORMATION

The final activity in the building of the DB is to take the 'joint' derived object summaries set out in the business study to define the *ad hoc* information requirements of management.

Each of the summaries should be examined to see if it should be included as a derived object or class of object. For instance, if a routine requirement exists for details of sales of product-type by industry, a judgement should be made whether this information should be contained in the industry file, or in the product-type file, or in a separate statistical file, or whether it should be generated on demand.

There is no standard answer but some of the considerations are as follows:

1. Where the same information is required for more than one type of derived object, the team should consider whether the information should be stored at the primary object level. If sales of *product-types* are required both by area and by industry it might make sense to maintain totals of area sales and industry type sales at the level of the *individual product*.
2. On large files there is a performance overhead in including masses of summarized information in object files. Conversely, on large files the response time will be great if virtual field summaries have to be created on demand. A partial answer is to create a separate summary file, especially if

it is acceptable to update it periodically in batch mode.

3. For *ad hoc* batch requests the best general answer is to create a single dustbin file with skeleton information for each individual activity. This file can be designed to accept every activity and then 'selected, sorted, and listed' to produce *ad hoc* statistics.
4. A summary total is rarely worth maintaining on file unless it is a summary of at least a dozen fields at a lower level. If there are only three or four sales areas per region it is probably not correct to maintain summaries at the regional level. As there are perhaps hundreds of customers per area it is probably worth maintaining summaries at the area level.
5. If the demand for *ad hoc* information is only a few dozen enquiries per hour, it is rarely worth maintaining any summaries except those which eliminate unacceptable response times. One eliminates this by the maintaining of 'running total' summaries as a by-product of each transaction.

If the demand for information is several hundred enquiries per hour on large files it is probably essential to analyse the demands in order to ascertain which intermediate totals might reduce the machine load.

Chapter 11

A Variation on the RDB

The concept advanced in this chapter is new and of interest primarily to the DP professional. In most DP applications the analyst meets cases where for ease of development or for fast response time a logical file which breaks the rules of the RDB could provide an obvious answer. He is constrained from developing a design to include such a file because of his awareness of the potential long-term danger. In many cases this danger is minute or nonexistent – if the file is an activity file.

BENEFITS OF DISTINGUISHING OBJECTS FROM ACTIVITIES

The traditional DP model makes no distinction between objects, groups or classes of objects, and activities in the real world. They are all entities and are treated identically in the process of analysis and design.

A shortcoming of this failing to distinguish between activities and objects emerges in obtaining information requirements from management. Managers can never define their information needs if the question is asked about procedures and activities but invariably give complete and consistent answers if interrogated about individual objects and groups of objects.

Another benefit of distinguishing between objects and activities is the ability to focus the aim on activity files when attempting to simplify the logical schema of the database as discussed in the previous chapter.

A further benefit of distinguishing between objects and activities is that it becomes possible to check that the data in the DB resides in the logical file demanded by the application as well as by relational theory. By knowing that most attributes of an activity are in fact the attributes of its associated primary objects it is easy to double-check the logical record contents of activities and their relation with objects.

A benefit arising from the distinction made this time between primary objects and derived objects is the check to ensure that data analysis at the primary object level is consistent with information requirements obtained in the first stages of the application in the object lists. In other words if putting the data from the primary object into 3NF does not indicate derived objects to meet all the information requirements outlined by management in the business study, then an omission has been made in the design of the new application.

Note that the normalization of data is carried out jointly from object listings, message content lists and DD entries. If it were merely message data which were normalized, as is assumed in many methodologies, this check would not be possible. Whenever a chance occurs to relate development activity right back to the business study documentation it should be taken.

A final benefit which is amplified in this chapter is the base that the distinction provides for *optionally* excluding activity files from the corporate DB. The reason for this is that it is common sense to take into account the usage characteristics of data, as well as its logical relationships, when specifying a logical schema.

This is often desirable because it results in a more efficient application in terms of machine resources, it can shorten the development time and it reduces the potential size of the central RDB and the overheads of maintaining it.

BENEFITS OF INTEGRATION VIA RDBs

One benefit of using an RDB is that the same logical data structure can be used later for applications that were not even thought about when the RDB was established. The RDB thus becomes a major mechanism for the integration of different applications – this integration being an unstated but ever-present design requirement for all applications.

Eventually the RDB will become a corporate DB serving as the base for information management through the entire organization. But if everything is considered as an entity to be inserted into this DB the structure will become enormous, as there are probably between 30 and 80 separate entities or relations in the majority of DP applications.

PROBLEMS OF SCALE

The problem of size has been raised by some of the more practical innovators, and the idea has been floated of major or significant entity types, of which there are perhaps 200 in an organization. One benefit of formally distinguishing between data design for activities and objects is that automatically the objects become the equivalent of these significant entities.

It is possible to make the distinction in design because the data which is associated with activities behaves differently to data which is associated with objects. Activity data also serves a different purpose within the macro-information system of the organization.

CHARACTERISTICS OF OBJECT AND ACTIVITY DATA

If the list of customers as it currently exists is compared with the list of customers as it existed a year ago, it would overlap to the extent that it would be almost identical. Conversely, if the list of customer orders is compared with the list of a year ago the two lists would be completely different (except in a few cases for the construction of high value capital goods).

Object files are always stable and activity files are always volatile as regards the records which they contain. Many customers remain on file for decades but few orders remain on file for more than a few weeks, or in special cases for more than a few months. In fact in the cases where the order is long lived, say for two or three years, the business analyst will often define the order as an object just because of its longevity.

This happens automatically during the business study when management indicate quite clearly that they have information requirements about each individual order. The order to them represents an entity which is associated with many activities, e.g. inspections on the supplier site, partial deliveries, financing operations, etc. The order becomes in effect a 'point of sale' entity.

Once one considers the individual data fields in the logical records, the opposite characteristic is true for both activities and objects. The attributes of an activity are much less likely to change during its life in the system than are the attributes of individual and group objects. An order may be cancelled but it is not often modified. If orders are frequently modified there is a case for distinguishing between 'letters of intent or provisional orders' and 'firm orders' during the business study. The labour balance will highlight what is happening. In this way the MIS application, which will eventually be constructed, will provide more meaningful figures.

If a high volume of modifications is allowed on orders or on other activities the system generally is not very robust, nor is it proof against clerical error or fraud.

Object records however are always being modified. Group objects are very probably modified automatically by the system each time a transaction is processed, and the same applies to individual object records. Each time that there is an activity there will be a modification to some of the derived information associated with objects, e.g. total sales to date, credit limit, area sales totals, etc. The activity file is a volatile file of fairly stable records and the object file is a stable file of volatile records.

Record utilization

Activity records generally exist only in one application or in very closely related applications. Information pertaining to a dispatch note is likely to be of interest only to the order-book and perhaps to the stock ledger applications. An invoice record is likely to be of use only in an invoicing program and the associated accounts receivables. An order record will be of interest generally only to the order-book application. Information about orders will be passed to further applications via the update of a group object, e.g. value of outstanding orders by industry.

Object files however are used across the entire organization. Most departments in a commercial organization will need to know something about the customer. Accounts, sales, marketing, production and distribution will all need to know something about the individual customer and about groups of customers.

The object record is always used across the organization but the activity record is generally used by only one application.

Record characteristics

A corollary is that as time progresses and more applications are developed, the object record becomes larger as additional attributes are added – and it was probably a large record to begin with. Contrarily, the activity record is small. An order usually contains the following.

1. Order ID.
2. Customer ID.
3. Product ID.
4. Quantity ordered.
5. Order date.

Everything else can be inferred from the attributes of the associated objects. Even if object attributes are duplicated into activity records, the resultant record is often only a few hundred characters in length.

The object file generally has large records, the activity file should contain small records.

THE INHERENT DATA STRUCTURE OF ACTIVITIES

However, there is nearly always a natural hierarchy of information in activity files which reflects the structure of the information outside the computer kernel. In the case of an order the hierarchy will be as follows:

1. Order header, contains customer details.
2. Order line (many to one header), contains product details.
3. Delivery header (many to one order line) contains delivery details.

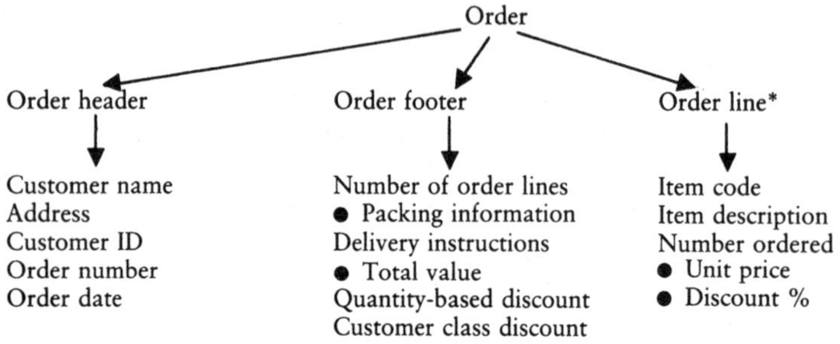

* There are many lines to one order. The maximum required number of lines per order is established during systems analysis.

Fig. 11.1 An inherent data structure.

There is an inherent structure to the data in an activity file which reflects the way in which it is used outside the computer kernel and this is called the 'inherent data structure' (see Fig. 11.1).

An object file has no inherent data structure, an activity file always has an inherent data structure.

Finally the usage of the information pertaining to individual activities and to individual objects is different. The raw information from an activity file is rarely used by more than a handful of people. The raw information pertaining to objects is rarely used by less then two or three different departments.

Information pertaining to individual activities has a limited distribution, that pertaining to individual and group objects has a wide distribution.

The summary of the differences between activity data and object data is summarized in Table 11.1.

ACTIVITY RECORDS: RDB AND 3NF

The aims of RDB and 3NF data analysis included the structuring of data so that it could be used easily for any purpose. The logical file for customers set

Table 11.1 Differences between activity data and object data

Object	Activity
Stable file	Volatile file
Volatile records	Stable records
No inherent data structure	Inherent data structure
Wide information usage	Limited information usage
Wide DP usage	Very limited DP usage
Large records	Small records
Long-life records	Short-life records

up for the order-book can be used unchanged or can be easily extended for accounts receivable, distribution, marketing and the MIS applications which are constructed on top of them. But, as already stated, the raw data in an activity file will not be needed for other purposes except via the provision of derived attributes to objects. Any non-RDB structure for activity files is unlikely ever to have implications for other applications.

There is no need for RDBs to provide flexibility of usage for activity logical records because in the real world it is not needed.

A second reason for using 3NF data is that if data changes it only has to be updated in one place in the DB. However, the activity record has such a short life that it invariably is eliminated from the computer system before it needs to be changed because an attribute has changed on an associated object. If a customer intends to change his address he will give plenty of notice. It is quite safe then to write the address into each logical order record because the order will be dead long before the address is required to be changed.

There is no need for 3NF to ensure data integrity (i.e. that all orders have the correct address and are for a current product) because an activity record ceases to exist before it need be updated.

There is no reason to eliminate repeating fields in an activity record. They have to be eliminated in an RDB because the RDB operates on fixed tables or matrixes whereas a simple file handler can handle variable numbers of fields in the same record, i.e. the situation where an order can have many items and an item can have many deliveries.

In nearly every case the inherent data structure of an order can easily be placed in a single record. This implies that variable length records are necessary, to store different numbers of order lines on different orders.

Activity files tend to have small records and an inherent data structure can be easily depicted in a single record if so required.

ADVANTAGES OF EXCLUDING THE ACTIVITY FILE FROM THE DB

1. The database controller does not have to be concerned with the allocation and structure of data elements in an activity file, except as they update object records in the DB.
2. The development process needs to involve the data management section in the MIS department only minimally.
3. The overheads of controlling the corporate DB are reduced.
4. The development team has more freedom to consider different approaches on structuring activities because the resultant logical files are unlikely to be inconsistent with the central corporate DB.
5. The application is likely to be faster than with an RDB. RDBs are becoming as fast as any other method of data management, but in the last

Order ID
Customer name*
Address*
Customer ID
Order date
Number of lines on order
Packing information
Delivery instructions
Order value
Quantity discount
Customer class discount*

Order items can be recorded using one of four methods:

1. A separate record is raised for each order item. If these records are included in the same logical file as the rest of the order information then the file becomes difficult to use for anything except order processing. In many cases this is merely a theoretical objection.
2. A separate order item logical file is created which is related to the order record. This is the accepted method but it imposes performance penalties.
3. Each line is allocated a set space within an order record which is always capable of handling the maximum permitted lines per order. This solution again is consistent with relational theory but imposes a penalty if the physical file handler does not automatically condense data.
4. Variable length logical records are stipulated to the physical file handler (if it is capable of handling them). This method usually provides the most efficient use of computer resources but results in the most inflexible file layout. In some cases this does not matter (e.g. if the order is processed and archived in a few hours).

* These fields are usually not entered at the time that the order is submitted to the computer. They may be extracted from the relevant primary object files in the relational database as this cuts the time taken to input the order and reduces the chance of clerical error. However, it is a valid design option to include the fields in the order record as it is an activity record and there is little danger of the information becoming out of date before it is used.

Fig. 11.2 The order-book record.

resort the construction of a user-view from a logical schema in an RDB demands on average two or three extra record retrievals.

This can be a bottleneck in high volume applications which process scores of transactions per second. For instance only a single record has to be accessed to retrieve all information for an order using the inherent data structure for the order-book file. If the data were in 3NF in the RDB the logical files would be as follows:

1. Order-book.
2. Customer.
3. Product (two or three times).
4. Probably two or three derived object files.
5. Postcode file.

The procedure then is to consider whether the activity file at the centre of the application can be *safely* excluded from the RDB as regards processing. The information to ascertain this is contained in the business analysis and particularly in the object listings, the labour balance and message flow charts.

If it can be excluded it should be. The combination of an inherent data structure in an activity file in conjuction with the concept of commonizing 'status-modification' messages on that file, as discussed in the previous chapter, will result in an application with a high performance requiring few programs. However, it is unsafe to make this decision without the application-oriented deliverables defined in this book to back it.

Chapter 12

The Program Function List: Kinds of Programs and Screens

NEED FOR STRUCTURED APPROACH TO ANALYSIS

Up to this stage the team has defined the business and operational framework of the applicaiton and the data structure in 3NF to reflect the business and operational requirements of the application. The team should also have decided on a centralized or distributed overall design for the application and should have decided whether to seek an application package or to write customized programs in a 3GL or a 4GL.

Whatever options are chosen during development, data analysis and design needs to be carried out. The use of object lists as a check document and the use of message content lists as prime input is a means of orienting the output to an operational and business view of the application. The guiding principle in design should be to make the computer system fit into the overall application and not vice versa.

This means that the application DB should be defined in broad-brush logical terms before a choice is made about application packages (e.g. the major entities in traditional development terms or the object logical files in the terms outlined in this book).

If this is not done there will be difficulties in siting the man–machine interface in the optimum place. There will probably also be difficulties in integrating the package with associated applications. The programs which mesh with manual activities in the application also should be listed and roughly classified before available packages are evaluated.

THE PROGRAM FUNCTION LIST

The final major stage in specifying the application is expanding the abstract definition of the required computer system which has been established

Message. Add new industry records.

1. Scan industry file for duplicate record. Reject message if duplicate
2. Scan industry grouping for duplicate record. Reject message if new industry grouping data for duplicate record.
3. Scan industry class for duplicate record. Reject message if new industry class data for duplicate record.
4. Syntax checks on industry name, group name, class name. Reject if checks fail.
5. Scan industry file for duplicate name. Reject if found.
6. Scan group file for duplicate name. Reject if found.
7. Scan class file for duplicate name. Reject if found.
8. Create new industry record.
9. Create new grouping record if any new grouping data.
10. Create new class record if any new class data.

Fig. 12.1 Program function list.

(messages and logical DB) to include a list of required programs. The programs may be written in-house, or purchased, or they may be versions of programs that exist elsewhere.

The list of required programs (the program function list) is the final step in defining the logical structure of the application. The program function list is not the same as program specifications. The list states what the programs should do logically, whereas program specifications deal with the mechanics of handling data in a specific DB or file handler on a specific machine.

The program function list is derived from the revised message contents list resulting from the analysis and comparison of message contents in the original list. For each message, the program function list notes the characteristics of the programs that are needed to connect the message to the logical files in the DB. Figure 12.1 gives an example of a program function list.

The program function list is an important project control document. It provides, with the logical schema of the DB, a basis for project estimating in terms of time and cost. It also allows the establishment of an application 'module library' to provide elements to be copied into 3GL application programs.

CATEGORIZING PROGRAMS

Categorizing the programs makes it possible to check that the specification process up to this point has produced a program list which meets the needs of the business. It is a document which links the computer-oriented specification back once more to the application-oriented specification behind it. The types of programs in both DP and MIS are fortunately not very large.

SUMMARY OF THE FUNCTIONAL CHARACTERISTICS OF PROGRAMS

1. Batch or transaction processing
2. Enquiry, and/or
3. Update, and/or
4. Insert new record, and/or
5. Delete record, and/or
6. Produce output, and/or
7. Validation required, and/or
8. Parameter input.

Batch or single transaction

It might be desirable sometimes to produce an invoice as the final result of processing a returned delivery note. The details of the completed delivery note are entered into the computer and a few seconds later an invoice is printed. If the volumes are low this transaction processing of the invoice might well be the optimum solution. But there are some disadvantages.

The invoice is printed on special stationery and the choice is between having a printer permanently loaded with invoicing stationery or loading the printer with the invoice paper every time an invoice is required.

An alternative is to produce an invoice record on disc, and to print all the invoices in one batch at the end of the working day.

Batch processing is generally used where there are high volumes of data to be processed. A good example is the production of gas or electricity bills (see Fig. 12.2).

In this case both the computer processing and the printing are carried out in batch. The process is more efficient in terms of machine utilization than is transaction processing (which is why the old 1960 mainframes were able to run massive applications with less computer power than a present day small micro) but is more inflexible.

Batch processing tends to be appropriate in the following cases:

1. Where there are high volumes.
2. Where most object logical records in a file are processed.
3. Where the processing cycle is regular.
4. Where the work of the manual side of the application falls naturally into a batch mode.

Input programs to provide data validation

Data validation is so important that it should be analysed as a separate activity. The aims should be as follows:

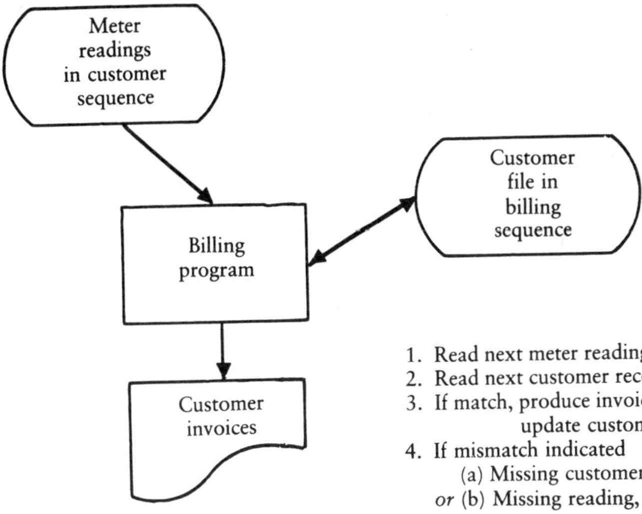

1. Read next meter reading.
2. Read next customer record.
3. If match, produce invoice.
   update customer record.
4. If mismatch indicated
   (a) Missing customer, go to 2.
   or (b) Missing reading, read next meter (go to 3).

Fig. 12.2 A simplified batch update.

1. That no incorrect indicative data should enter the system and
2. That no unlikely quantitative data should enter the system.

   A balance should be struck on the validation of quantitative data because excessive validation sometimes places an unacceptable burden on the end-user.

Validation programs should be noted so that the insertion of validation into the application can be considered from this stage. Validation has to be incorporated into the architecture of the application, especially if it is a financial system, rather than applied as make-up to cover blemishes of design.

The functions of the program, e.g. whether the program changes or merely interrogates the DB

Where possible the same screen should be used for inserting, deleting, interrogating and modifying records to reduce programming and the number of end-user screens. The program function list allows this to be checked for single and multiple file processing from a single screen.

Entering parameters for another program

An obvious example is the entering of parameters to generate a print report. All the program functions in the application theoretically can be called into operation directly by a program or indirectly by a set of parameters from a parameter file.

The parameters can be provided by the end-user, so that the extent to which this type of program is used effectively defines the extent of end-user programming or 'Do It Yourself' processing. In fact, end-user 4GLs are usually no more than a parameter file linked to powerful macro instructions to process them.

If a development option indicated that the maximum operational use was to be made of end-user computing, the program function list provides a base from which the analysis for fixing the extent may be made against a yardstick (e.g. interrogation only, can be easily parameterized).

SCREEN TYPES

The screens are the only part of the system that the user sees. To the end-user the screens are the application. Some common faults in screen design are detailed below.

1. Screens contain too much information, they are 'too busy'. Busy screens are not a great disadvantage when carrying information on given attributes for objects. The given attributes of objects rarely change and there are usually a lot of them.

In the trade-off between a few procedures with few complex screens, and many procedures with simple screens, the design thrust for the given attributes of objects should be to minimize the number of screens. They will not be used very often.

2. Conversely a transaction screen should carry the absolute minimum of information. One might have a common screen say for inserting/modifying/deleting/interrogating the name, address, delivery address, telephone no., contact name and telephone no., terms of trade, industry code, etc., of a customer. It might even be the case that there would need to be coupled screens because one screen could not carry all the information.

However, it might be bad design to have a common screen for order enquiry and order entry even if the two were practically identical. If both screens are in continual use they need to be stripped for action. An order screen needs a delivery address, either typed in by the user or provided from the customer logical file; an enquiry screen does not.

Similarly it might be incorrect to have a common screen for order entry and raising an advice of delivery note. A screen for advice of delivery might be as follows:

1. Type in the order number ID.
2. The computer generates the customer name and the product names.
3. The operator types in the quantity dispatched.
4. The computer gives a warning if the quantity does not match the original order quantity.

There is no address information, no customer code, product codes, delivery instructions etc.

The invoice input screen might be as follows:

1. Type in the returned delivery note number and the customer code. Note a financial transaction that gets no further if the computer detects a mismatch.
2. The computer generates an invoice screen with warning for mismatched quantities.
3. The invoice record can be optionally changed (keys apart) before being placed into a file for subsequent printing.

A transaction screen should need minimal input from the end-user and contain information only for that transaction.

3. The method of screen management does not match user skills and processing speed. When selecting an option from a menu, the choices are either to type the option number into a communication area or to position the cursor onto the menu. The user should be given the method he already knows from other applications.

On completion of the activity, the next action may be chosen either by returning automatically to the previous menu screen, or by pressing 'help', or by selecting from a mini-menu at the bottom of the screen. Whatever the user is habituated to should become the standard.

However, the user should never be required laboriously to navigate a structure of menus to select the next screen. It is desirable in a system of hierarchical menus to permit the user to nominate his next screen directly in a communication field and to use the menu system only as a default option.

4. The entry of data to the screen should match operator manual activities. For instance data should be depicted on the screen in the order in which it comes to the operator's eye. Screens should never need data entry in a certain sequence; if the user wants to grasshopper his entries he should be allowed to do so.

5. There should be one screen for one document (although many documents might use the same screen). The user does not want to flick through a pile of documents to enter the next field.

6. There should always be an effective 'help' button. The messages it provides should be examined after a few months of system operation and changed.

7. If text is entered it should be possible to correct the text as with a WP. It should not be necessary to erase the entire field entry and retype. Many screen painters and screen managers do not offer this feature and its lack can be a major worry where extensive text has to be typed onto structured documents.

Examples are special make instructions, special delivery instructions, technical specifications, etc.

8. In most applications, especially for intermittent users, the use of Function keys should be avoided. The ideal keyboard is no more complicated than that of a typewriter.

9. Try to ease comprehension by highlighting the line where the cursor resides. This eases data entry and retrieval on busy screens.

10. Never design a code field of more than eight characters. If necessary make it into two separate fields on both source document and screen.

COSTING THE PROGRAMS

It is now possible to cost the programming accurately from the program function list and the list of screens, divided into the busy screen for object information and the lightweight screen for processing operational transactions.

The simplest method is to establish a fixed cost per program, to be added to a fixed cost for each function performed by the program, to be added in turn to a fixed cost for busy and lightweight screens respectively.

The traditional method of estimating lines of code and assuming that so many lines can be produced per programming hour never was very good. The variation between methods in common use often gave answers varying by 3:1 for the same programs. The above method is simple and it provides good answers.

Chapter 13

Systems Analysis and Fourth Generation Technology

The procedures outlined in the book have been illustrated largely with respect to a traditional approach to a traditional problem, i.e. developing a customized system using traditional 'successive definition stages' where each stage takes the deliverable documentation of the previous stage and expands it to include more detail. Only at the end of this process are the programs constructed.

Unless this systematic process of analysis and specification is carried out the individual programs in the application will not fit together. They will then have to be modified or rewritten, and when they do eventually communicate with each other they will invariably omit many key activities from the user viewpoint. This painful lesson was learned in the 1960s.

Unfortunately the writing and rewriting of a 3GL program is a time-consuming and expensive operation, so expensive indeed that the use of structured methods can be justified merely be reducing the chance that it might occur. 3GL programming costs can sink any project unless they are tightly controlled.

3GL programming is so expensive and time-consuming that the writing of completely new customized programs has been an activity of last resort for some years for experienced DP professionals. There has been an increasing use of application packages, and the modification of programs which have already been developed on other applications, and finally there has been the advent of 4GT where if customized programs must be written the process is so fast and inexpensive that an error in specification is no longer fatal to the entire project.

A crucial question is whether the procedures outlined in this book can be applied to the selection of packages and the use of fourth-generation techniques, which now probably account for most new business applications. If traditional structured analysis methods exist only because they represent a way to control projects with very high programming costs, then it is likely that they would not be necessary for developing systems with low programming costs (i.e. application packages and 4GT). However, the control of expensive and time-consuming programming is not the only reason for structured development (i.e. formal sequence, deliverable documentation, top-down).

It is worth recapping the development process at this stage to see the relevance to 4GT.

SUMMARY OF THE METHOD

1. Fix objectives and justify the project – wish-lists and mission statement.
2. Fix the boundaries of the application – DFDs and activity chain.
3. Define information requirements, broad brush and then in more detail – object lists.
4. Think about a feasibility study.
5. Analyse the existing system (if appropriate) – message lists, labour balance, message flow chart.
6. Define required overall application in information terms – application block diagram.
7. Consider the options.
8. Split into manual and computer components – block diagrams.
9. Start thinking about required messages – message contents lists, activity chains.
10. Build a data directory.
11. Build the logical schema – logical file lists.
12. Simplify the schema.
13. Commonize finally all messages.
14. List required programs – program function list.
15. List screens – screen list.

In traditional development it is well after this stage that program specifications are produced and the actual construction of the programs takes place. Apart from the design errors which are generated by designing the new system before business and operational requirements are fully established, and before a man–machine interface has been fixed based on operational requirements, the unavoidable weakness of traditional systems design is that users and management must agree the operational system in advance by scrutinizing a lot of complicated documents. They always have to agree a system before they have had any exposure to it, rather like buying a car without the benefit of a test drive.

The use of 4GT can provide at least a limited test drive by showing working screens or even a working prototype to users, because 4GT programming demands only about 10 per cent of the cost and perhaps 5 per cent of the time of traditional programming. This seems to stand the development process on its head, because programs can often be written while major design decisions still have to be taken. That is a radical idea. For instance, programs might be written as early as activity 3 (establish end-user requirements) and long before activity 14 had even been considered (establish a program function list).

User-oriented deliverable documentation eliminates some traditional documentation, orients the remainder to the user and manager, and involves the user more in the design process than is allowed by the documentation of traditional technology-led design. However, there will still be problems in meeting operational expectations, because at the end of the day the user still has to agree a system before he or she has had any exposure to it operationally.

For instance, providing the necessary response time in a telephone ordering system is very difficult until experience has been gathered of the kind of supplementary questions that the customer will ask while deciding on his order. If the designer tries to provide for everything he will provide a system that is too complicated to use.

The idea has grown in some 4GT circles that defining everything about the application in advance of programming is no longer necessary. If the system is not acceptable another set of programs can always be knocked up in a hurry. In other words one can proceed by building approximate prototypes of the application and refine them in the light of user feedback.

However, the argument is sometimes overstated. The principal stages for developing systems remain as they were for 3GT if the application is to be based on a detailed appreciation of the operational and business needs. The user can be led to define the optimum application only by someone who has taken the trouble to uncover the nature of the application.

Activities 1–10 still need to be carried out before the construction of the system is started, else quite simply the wrong system might be constructed.

Business analysis and operational analysis must both be done before designing the computer system, no matter how easily a computer system can be written. If they are omitted the door stays open to the facile design and dismissal of the messier operational requirements that were the real bane of 3GL systems.

Similarly in the construction of the computer system, RDB and 3NF data remain the only robust way of organizing information inside the computer. Therefore the same data analysis has to be carried out as before for any application with more than a dozen or so logical files.

Program function lists also need to be derived on the basis of an analysis of requirements rather than the heuristic method of 'prototyping until it's right'. And function lists need to be established because without them and screen lists it is impossible to cost program development or to estimate operational costs. Activities 11–14 still need to be carried out.

There is a further point. One purpose of development documentation is to allow an analysis to be made before implementation to ensure that the functionality (the procedures) and the performance of the system meet the requirements laid out in the business study.

It is not acceptable to produce a working system based on user requirements as they emerge during the prototyping process and then leave it at that; the system might not be able to cope with something that comes up tomorrow for the first time. Without documentation it is impossible to say whether this may be the case.

Additionally, a computer system cannot be tested over all the permutations of the data which are processed and so can never be 'proved' operationally. A system has to be 'proved' by an analysis of the documentation defining its structure.

Theoretically this is impossible at the present state of the art, but at least one can spot the majority of things that are likely to go wrong, or go too slowly. Not only does a 4GT application need to go through the stages and activities of business analysis, it also has to produce the same kind of documentation so that the system can be analysed and checked out logically.

It may be that the stages of development do not contain the same level of detailed documentation needed for 3GT applications, because much of the detail is needed for producing consistent program specifications which are surplus to requirements with 4GT. But the structured stages must be percoursed and must result in deliverable documentation – but when, exactly?

Most 4GT produces documentation from the data directory (DD) of the application, but the DD is often used first as the base for a working prototype, and only later to produce system documentation. Activity 15 could start immediately after activity 2; activity 9 could precede 6.

Outside the 4GT world, it is not fully understood that there are many different kinds of prototyping which are quite different in their development methods. Examples are as follows:

1. Iterative prototyping, i.e. mark (1) through to mark (*n*).
2. Incremental prototyping, i.e. start with the core and work outwards.
3. Interface prototyping, e.g. screens and reports.
4. True prototyping, i.e. scrapped and rewritten in a 3GL.
5. As above, but save the DD.
6. Pseudo-prototyping, i.e. the 'prototype' becomes the working application.
7. Perm from the above.

All of them (and their variants) *begin* the structured stages of development in different sequences. But they all have to *terminate* structured stages in the same sequence and that sequence is the same as it was with 3GT. The bottom line is the same as it always was (deliverable control documents produced top-down), but the ways to reach it are various.

This seems to be the crux of applying any development methodology to 4GT applications. Activities which are rightly required in a certain sequence in 3GT should be presented so that they can be controlled in the different starting sequences demanded by the different prototyping approaches. There is no danger of losing control, as all prototyping stresses or simply assumes the need to control the DD. Where there is no DD there is invariably tight 'change control'.

There is nothing wrong with prototyping. If a prototype prompts object listings from management to establish information requirements, the net result will be no different to obtaining the object lists in the traditional fashion. The important point is that object listings shall be produced before very much money is expended, and that they should be used to help provide a business oriented definition of the application.

This book indicates activities which have to be carried out in business analysis and system design regardless of the technology used. It also lists documentation of the type which must be produced to indicate to computer management that the system meets technical requirements. It does not matter if the deliverable documentation is produced in serial fashion or if the activities are initiated out of sequence to make full use of 4GT. *But the stages have to be signed off in the sequence in which they have been presented.*

In other words, even if prototyping is used a logical schema for a DB cannot be signed off before the object lists have been agreed. The logical schema might be used to trigger the production of the object lists, but the schema cannot be finally agreed until the object lists and associated early stage deliverables have been signed off. If documented applications are produced with 4GT, and the documents are used to fix essential development activities, the use of 4GT, although informal and nonbureaucratic, implies tight technical control.

Prototyping is often seen as incompatible with structured methods. In fact, successful prototyping always demands a highly structured approach. And a structured approach can always be based on business oriented definitions.

The reluctance of prototypers to use traditional structured methods is largely because both sides misunderstand each other's viewpoint. Structured methodologies contain techniques and procedures which are of help to most forms of prototyping.

Eyeball observation for instance perceives the following strategic control activities for different kinds of prototyping:

1. Iterative. Functional definition of successive versions (e.g. via activity chains).
2. Incremental. Data consistency of core and periphery (e.g. via object listings).
3. Interface. Change control.
4. Truc. Procedural analysis (e.g. via message flow charts).
5. True with DD saved. Ditto.

6. Pseudo. Data structuring. De-bottlenecking (e.g. via message content lists, commonizing messages).

PACKAGES

Most people select packages too soon in the development process. The most common route seems to be to seek out packages as soon as a name is given to an application (e.g. general ledger, inventory control) and then to try to decide which package will fit best into the application technical environment. The key to the entire project then becomes the most efficient use of the package rather than meeting user requirements.

The converse should apply. Even if the chosen option is to procure a package to modify in-house, the business analysis should first be carried out. Nor should the decision be made to procure a package before the process of examining options has been carried out, and this implies an analysis of the application area. This is rarely done, as in many installations it never occurs to the computer people that such an analysis is needed.

This is usually why wrong options are chosen, such as providing complicated reports instead of a simple report-writer for the end-user; or the decision to develop a mainframe system when networked PCs are better suited to user needs; or the assumption that the application is wholly centralized or decentralized, when there is a mix of centralized and decentralized activities.

For both packages and productivity tools the evaluation of options should be carried to the stage of examining possible alternatives on the basis of the labour balance. Only when this stage has been reached in the project is it possible to consider the use of packages and productivity tools instead of a custom-built application.

Business analysis and application design based on the detail of the real world is common to all business applications and is needed regardless of the technology of development and the technical option chosen. And finally it is needed to allow the systematic construction of an optimal application.

Glossary of Terms

activity A category of entity. The temporary association of two or more objects in performing an action, e.g. a sale, an order, a dispatch.
activity attributes A list of data elements for each activity file.
activity chain A series of connected messages and actions which constitute a major business activity. A typical activity chain can be conceived as beginning with the receipt of an enquiry and ending with the acceptance of the order by the customer.
activity files A list of all the logical files about activities.
added value A form of benefit resulting from a DP or MIS application. It represents additional earnings or revenues which are a direct or indirect consequence of the application.
analyst work bench (**AWB**) A set of productivity tools for the systems developer.
application package A set of programs designed for a particular type of application which can be customized to meet the needs of an individual firm, e.g. general ledger, materials management, production scheduling.
attribute Codified information about an entity.
batch processing Data processing via a series of partial processings of a batch of information, i.e. the entire batch is partially processed before any member of the batch is completely processed.
box diagram A diagram showing the information inputs and outputs of the application. Box diagrams may represent the total application area, or may represent the computer processing at the kernel of the application. Box diagrams as they are developed in more detail may also represent messages, as well as general statements of information flow.
CASE Computer aided software engineering. Many CASE tools are concerned with analysis and design and have little to do with software.
context diagram A simplified data flow diagram.
cost displacement A reduction in operational costs as a direct or indirect result of an application.
data dictionary (DD) A definition of the characteristics and the relationships of every data element in an application.
data flow diagram (DFD) A chart depicting the flow and filing of data in the application area.
database *see* DB.
DB A collection of cross-referenced computer files. There are three basic DB structures; hierarchical, networked, and relational.

DBMS A set of software used by application programs to manipulate a database. The DBMS manages physical records and files on disc, transforms physical records into the logical records utilized by application programs, and associates different logical records to construct records which reflect a user-view of the application.

direct benefits Added value or cost displacement benefits which accrue as a direct consequence of an application. Sometimes known as tangible benefits.

distributed systems 1. An application where control over the functionality of the application is decentralized to the individual user.
2. An application where computer resources are distributed throughout a network instead of being concentrated in one or two major points.

DP Data processing. The applications which capture and process the raw data of everyday administration, e.g. payroll, ledgers, invoicing, etc.

dustbin file An unsequenced file where individual records are retrieved by 'rummaging through the dustbin'.

entity DP terminology for anything about which it is necessary to gather information. The three kinds of entity discussed in this book are simple objects, classes of objects (i.e. conceptual entities) and activities.

entity life history A document used in structured analysis to ensure that all logical records will be established, maintained and deleted by proposed programs.

feasibility study A short preliminary business study for large applications.

field A location on a document, on a screen or in a file for the storage of a single data element.

financial system An application concerned with the flow of money through an organization.

fourth generation language (4GL) A language which enables programs to be written about an order of magnitude more quickly than does a traditional language.

fourth generation technology (4GT) Software which permits the user to state requirements in terms of the problem rather than in terms of the computer, e.g. Spreadsheets, 4GLs, electronic mail, etc.

given attribute Information about an entity which is submitted to the computer system.

green field application An application in an area where previously there was no application of any sort.

hierarchical database A database with a tree structure so that any record can only depend from a single higher level record in a given tree.

indirect benefits Benefits which are made possible by an application but are not an automatic consequence of the application. They generally occur elsewhere in the organization. They are sometimes known as intangible benefits.

inherent data structure The way that people think about the structure of information associated with an activity in the real world.

key indicative data Data which is simultaneously an attribute in one logical file and an entity or key in another logical file.

labour balance A matrix which relates the processing of messages to the people who spend time in the processing.

logical schema The organization of the data in a database as it is perceived by the systems designer. Data also has an organization as viewed by the computer operating system (the physical schema), and as viewed by the user of the application (a collection of user-views).

logistic system An application concerned with the flow of 'things' through an organization. Note that a 'thing' can be a service and in some cases can be money.

message A packet of information which travels from point to point in the application area.

message contents list A document giving information on the fields and the data elements contained on each message.

message flow chart A diagram for each major message type showing its movement from origin to destination and the processing carried out *en route*.

message list A list showing message name and type, average and peak volumes.

MIS Management Information System. The applications which accept raw data and processed information from DP applications to provide information for management. Good MIS applications are integrated with lower level applications.

mission statement An unambiguous statement of the expected consequences of the application in business terms. The mission statement gives in detail the 'object of the exercise'.

network database A database where records are in fixed hierarchies, but may depend from more than one higher level record in the same hierarchy.

object An entity which exists in the real world and which can carry out activities. An object is a person, a thing or a legal entity.

object attributes A list of all the data elements for each object file.

object class A group of objects which has its own set of unique attributes.

object files A list of all the logical files about objects.

object list(ing)s 1. A list of objects.
2. A list of the attributes of objects.
3. A list of object classes (conceptual groupings of objects).

processing cycle The calender time (usually daily, weekly or monthly) between the batch processing of individual transactions.

project steering committee An essential mechanism for management guidance and control during application development.

prototyping A group of development methodologies in which it is unusual to write formal program specifications before programming is started. There are many different kinds of prototyping.

relational database (RDB) A database without fixed hierarchies of records and in which all the attributes contained in a logical record refer to the entire key of the record (and not just a portion of it), and nothing else.

Report-writer A program for constructing reports on the basis of parameters provided by the user.

structured development 'Structured' in Dp usage usually means 'systematic' and structured development is systematic development where one activity is formally terminated before a subsequent activity is started. A rigid series of procedures and documentation which is systematically followed during development.

systems analysis The specification of the business and detailed operational requirements of a proposed application. The analysis activity processes an initial series of broad-brush statements into more detailed statements of their operational consequences.

systems design The design of a data structure and procedures to meet the business and operational requirements laid out in systems analysis. Note that analysis (what should be done) shades into design (how it should be done) as it is impossible to separate the two questions.

system generated attribute Information about an entity which is generated by the computer system, usually as a result of processing a transaction or of summarizing and analysing transactions.

3GL A third generation language such as Cobol, Fortran. 3GLs are high level languages which describe *how* program logic should operate. They are 'procedural languages' as opposed to 'declarative languages' which merely state *what* the program is required to produce. Most fourth generation languages (4GLs) are declarative.

third normal form (3NF) A data structure where every data element is written once only and retrieved by cross-indexing when it is required.

transaction processing Processing in which a single transaction is completely processed before any processing is allowed in the same program on subsequent transactions.

turnaround document A computer-produced document for subsequent input to the system at a later stage in the activity chain.

wish-list A statement giving an individual perception of the ranking and relative importance of the business objectives of the application.

YTD Year to date. Usually describes rolling totals of particular activities, e.g. Sales YTD.

Index